Human Minds and Animal Stories

The power of stories to raise our concern for animals has been postulated throughout history by countless scholars, activists, and writers, including such greats as Thomas Hardy and Leo Tolstoy. This is the first book to investigate that power and explain the psychological and cultural mechanisms behind it. It does so by presenting the results of an experimental project that involved thousands of participants, texts representing various genres and national literatures, and the cooperation of an internationally acclaimed bestselling author. Combining psychological research with insights from animal studies, ecocriticism, and other fields in the environmental humanities, the book not only provides evidence that animal stories can make us care for other species, but also shows that their effects are more complex and fascinating than we have ever thought. In this way, the book makes a groundbreaking contribution to the study of relations between literature and the nonhuman world as well as to the study of how literature changes our minds and society.

Wojciech Małecki is an Assistant Professor at the University of Wrocław, Poland. He specializes in literary theory, the environmental humanities, American pragmatism, aesthetics, and the empirical study of literature. He is the author and editor of five books and of numerous articles published in journals such as *The Oxford Literary Review*, *Poetics*, *Angelaki*, and *PLoS One*.

Piotr Sorokowski is an Associate Professor and head of the Department of Psychology at the University of Wrocław, Poland. He has published more than seventy research articles related to evolutionary, cultural, and social psychology, including in *Nature, Evolution and Human Behavior,* and *Journal of Cross-Cultural Psychology*. His work has been discussed by the media all over the world, including BBC, CNN, *Time*, and *The New Yorker*.

Bogusław Pawłowski is a Professor of biology and head of the Department of Human Biology at the University of Wrocław, Poland. He deals with human behavior and preferences in relationship to body morphology and physiology. He has published more than 80 papers in top journals in his field (e.g. in *Nature, PNAS, Proc. Roy. Soc. B., Current Anthropology*) and dozens of book chapters. He is the President of the Polish Society for Human and Evolution Studies (PTNCE).

Marcin Cieński is a Professor of literary history and comparative literature and the Dean of the Faculty of Philology at the University of Wrocław. His research interests include eighteenth-century and contemporary literature. He has authored and edited more than 150 publications, including *The Landscapes of the Enlightened; Polish Enlightenment Literature and the European Tradition;* and *Polish Humanism and Communities*.

Routledge Studies in World Literatures and the Environment

Captivity Literature and the Environment
Nineteenth-Century American Cross-Cultural Collaborations
Kyhl D. Lyndgaard

Ecogothic in Nineteenth-Century American Literature
Edited by Dawn Keetley and Matthew Wynn Sivils

The Ecophobia Hypothesis
Simon Estok

The Radical Ecology of the Shelleys
Eros and Environment
Colin Carman

Roads, Mobility, and Violence in Indigenous Literature and Art from North America
Deena Rymhs

Human Minds and Animal Stories
How Narratives Make Us Care About Other Species
Wojciech Małecki, Piotr Sorokowski, Bogusław Pawłowski, and Marcin Cieński

Human Minds and Animal Stories

How Narratives Make Us Care About Other Species

Wojciech Małecki,
Piotr Sorokowski,
Bogusław Pawłowski, and
Marcin Cieński

Routledge
Taylor & Francis Group
NEW YORK AND LONDON

An electronic version of this book is freely available, thanks
to the support of libraries working with Knowledge Unlatched
(KU). KU is a collaborative initiative designed to make
high quality books Open Access for the public good. The
Open Access ISBN for this book is 9780429061424. More
information about the initiative and links to the Open Access
version can be found at www.knowledgeunlatched.org.

First published 2019
by Routledge
52 Vanderbilt Avenue, New York, NY 10017

and by Routledge
2 Park Square, Milton Park, Abingdon, Oxon, OX14 4RN

First issued in paperback 2020

*Routledge is an imprint of the Taylor & Francis Group, an
informa business*

Library of Congress Cataloging-in-Publication Data
A catalog record for this title has been requested

ISBN 13: 978-0-367-66196-0 (pbk)
ISBN 13: 978-0-367-14604-7 (hbk)

DOI: 10.4324/9780429061424

Typeset in Sabon
by codeMantra

Contents

Acknowledgments   ix

Introduction   1

1   Texts, Statistics, and Deception   24

2   A Monkey, a Book, and Facebook, or How
    to Catch a Story in the Act   53

3   Does It Matter If It Is True?   69

4   Does It Matter How It Is Told?   85

5   Does It Matter Who It Is About?   110

6   How Does It Work?   128

7   How Long Will It Work?   145

    Conclusions, Speculations, and Prospects   153

Appendices   161
Index   183

Acknowledgments

The research presented in this book would not have been possible without the generous financial support we received from the National Science Center, Poland (grant number 2012/07/B/HS2/02278). We are grateful to the Center for believing in our project and for funding it.

While working on this project, we received help from various individuals, whom we would like to offer our thanks here. First of all, we greatly appreciate the cooperation of Marek Krajewski and his publisher ZNAK, which may well be the most extraordinary case of a bestselling author working together with scholars studying literature. It not only allowed us to obtain some extraordinary experimental results, but was also an adventure we will always remember.

Michał Kanonowicz and Agnieszka Sabiniewicz, our research assistants, helped us enormously with conducting our experiments. No words can describe the efforts they put into this, and they play an important part in the story of our project. We owe them special thanks indeed. Thanks go also to Elżbieta Bereza-Banach, who helped us considerably with various practical aspects of the project that would have otherwise been overwhelming, and Anna Oleszkiewicz, who performed some of the statistical analyses we used in Chapter 6.

While all these people helped us on the way that led to the writing of this book, others helped us in many ways when its first draft was already written. We are grateful in particular to David Wall, who was invaluable in assisting us with editing the manuscript. David offered countless suggestions on its style and content, and there is no doubt that the book would have been much worse without his help.

That the book eventually found its home at Routledge we owe to Scott Slovic and Swarnalatha Rangarajan, the editors of the series in which this book appears, and Michelle Salyga, our commissioning editor at Routledge. We would like thank them for their faith in this project and for supporting it on various stages of the publishing process. We are also grateful to the external reviewers of our book proposal for their insightful feedback, which helped us to improve the full manuscript. Also very helpful were the comments on the penultimate draft of the book that were given by Alexa Weik von Mossner. We owe her a great debt of gratitude. Finally, we would like to thank Harold Herzog, who not only read the final draft of the book, but was also kind enough to write an endorsement for it. This means a lot to us, especially given that this book was inspired in many ways by his own work on the psychology of human-animal relations.

We would also like to acknowledge that parts of Chapter 2 were previously published in the article "Literary Fiction Influences Attitudes Toward Animal Welfare" (*PLoS ONE* 11(12): e0168695), while some of the material included in Chapter 7 was taken from the paper "Can Fiction Make Us Kinder to Other Species?" published in the issue 66c (2018) of the journal *Poetics*. The respective parts of the book profited enormously from the comments made by the anonymous reviewers of these papers, as well as by our editors at *Poetics* and *PLoS One*, Shyon Baumann, Edna Hilmann, and Jennifer Lena, and by Andrzej Elżanowski, who read early drafts of these papers. We would like to use this occasion to thank them for their insights.

Introduction

The Humane Bestseller

In 1877, the English press Jarrold and Sons published a novel on the life and times of a horse. The author was a complete amateur, virtually unknown beforehand, but her book immediately appealed to thousands, then to millions, eventually becoming one of the biggest-sellers of all time. This, as anyone will admit, is a remarkable achievement in itself. But there is something still more remarkable about that novel. What we mean here is that shortly after publication, it generated a powerful wave of concern about equine welfare, which spread all over Great Britain and much of the Western world. Moved by the story of suffering inflicted upon its protagonist by callous humans, its nineteenth-century readers wrote angry letters of protest to newspapers, joined humane societies, and urged their political representatives to implement legal measures banning widespread forms of cruelty toward horses. Their efforts were successful. Some of those practices (such as "draining the blood of quarter horses to make them trot slower") were indeed banned, while some others (including the bearing rein) became less accepted.[1] And all this because of a single book!

That book was Anna Sewell's *Black Beauty* (2012 [1877]), and today, well over a hundred years after its publication, it is still popular among the general reading public. *Black Beauty* has also become a regular point of reference for animal advocates, as well as scholars from all kinds of fields devoted to human–animal relations, including animal studies, anthrozoology, and ecocriticism.[2] While those scholars and advocates invoke the novel for various reasons, one of those reasons is directly related to the topic of our book; *Black Beauty* may be seen as a perfect historical example of the power of narratives to shape the way we think about animals.[3]

Why Do Animal Advocates Want to Tell Stories?

The animal advocates' interest in the impact of narratives is related to the fact that despite all the progress made since the nineteenth century

DOI: 10.4324/9780429061424-1

in other areas, the systematic abuse of animals is still a common occurrence. Indeed, it has escalated to historically unprecedented levels, as exemplified by the current plight of billions of farm and laboratory animals (Singer and Mason 2006; Foer 2009; Singer 2009).[4] Since this exploitation would not be possible without the tacit consent of the wider public, a key task for animal advocates is to change the public's attitudes (Webster 2008, 19; Corbey and Lanjouw 2013). It is here where animal advocates think stories might help.

But why use stories for that purpose? Aren't there better instruments of persuasion? Wouldn't it be simpler to confront the public with the scientific data on the similarities between human and animal suffering (Dawkins 2008; Gregory 2008) and with arguments that make use of that data to argue that many of our current practices involving animals are morally wrong?

Unfortunately, the effectiveness of argumentative persuasion is limited. To understand this, consider the all-too-common experience of having clearly won an argument on a moral issue (be it animal experimentation, abortion, pornography, or helping refugees), yet at the same time having failed to convince our opponent. Probably, every reader has also had the experience of being convinced by an argument on a moral issue, yet not having taken its message to heart, and continuing to behave as if the argument was wrong nonetheless. Such cases are consistent with research in experimental psychology and philosophy that argues that moral attitudes (and attitudes toward the well-being of others belong to that category) are based on basic, unconscious intuitions that are highly resistant to being changed through argumentation (Rorty 1998, 167–201; Haidt 2001; cf. Aaltola 2010; Elżanowski 2013). Perhaps it is such resistance that underlies the fact that the situation of pigs, hens, and cows have improved relatively little in the recent decades. All that despite the steady growth of social awareness of arguments showing the immorality of current practices involved in factory farming (McGinn 1997, 207).

This is one reason why someone might turn to stories for this purpose, for another growing body of research indicates that moral intuitions often yield to *narrative* persuasion. This lesson is the staple of the so-called "narrative turn" that has been sweeping through academia and beyond,[5] a trend which involves, among others, all those journalists, media experts, and politicians who constantly abuse the term "narrative" and all those popular self-help books that present storytelling as an essential means of persuasion and a key to success in all kinds of pursuits (Salmon 2010). Perhaps it is. There are respectable studies which argue that narrative persuasion is useful in business, medical profession, and even in academia itself (O'Connor 2002; Kreuter et al. 2007). For instance, on the very day when this paragraph was written, an article was published in the journal *PLOS One* to the effect that those scholarly articles on climate change which involve a narrative are cited more frequently than those that do not! (Hillier, Kelly, and Klinger 2016)

But animal advocates were convinced about the persuasive power of stories well before anyone even began dreaming about taking a narrative turn. For instance, within the nineteenth century's culture of "sentimental liberalism" (Camfield 2005; Pearson 2011), stories were commonly believed to be an indispensable tool for improving attitudes toward animals, and they were systematically deployed for that purpose by writers, activists, educators, and organizations such as humane societies (Pollock 2005; Cosslett 2006; Boggs 2013; Eitler 2014; Davis 2016). Today, sentimental liberalism may no longer be part of our political culture, but pro-animal and environmental organizations still try to reach out to the public with the help of stories, including the heart-breaking narratives about the plight of abused horses, dogs, tigers, rhinos, and other animals that we know from advertisements published in newspapers and all over the Internet (Savvides 2013; Actman 2015; Greenfield 2016; Pijoos 2016; "Saving the Survivors" 2017).

Based on anecdotal evidence, they certainly seem to be on the right track. Consider for example the journalistic report "They Die Piece by Piece" on the slaughtering practices in the USA, published in *The Washington Post* in 2001. As the activist behind the story, Gail Eisnitz, recalls:

> the public's response [to it] was one of the highest reader response stories that they have ever done. Thousands and thousands of letters, e-mails and phone calls flooded in expressing gratitude and outrage — so in this sense it was a successful story. A lot has happened in conjunction with this story. One thing is that this story has had a tremendous impact on U.S. Congress and as a direct result of the story members of Congress and U.S. Senators were horrified with the information. That then enabled us to go to them and they introduced resolutions in the U.S. Congress demanding upgraded enforcements of the Humane Slaughter Act.
>
> ("Interview with Gail Eisnitz, Author of 'Slaughterhouse'" n.d.)

Or consider the famous case of the Tamworth Two, a pair of British pigs who ran away from "the truck taking them to the slaughter, burrowed under a fence, and fled into a thicket from which they could not be induced to come out" (Masson 2008, 24). Their story was widely covered by the media around the world to the applause of the cheering public (cheering for the animals, not those chasing them), which eventually helped to write its happy ending. Instead of going under the knife, they landed in an animal sanctuary (24). After all, who would want to send to slaughter the hero of a story one likes so much? In the light of this anecdote, we could even lend an ear to those who link the fact that in 1995 "the US Department of Agriculture showed stagnant demand for pork" to the huge popularity of the movie *Babe* at that time – the story of a piglet everybody seemed to have fallen in love with (O'Connor 1995, 16).[6]

Why Do Scholars Want to Study Animal Stories?

While scholars who study human–animal relations may be interested in narratives for the same reasons as those which the animal advocates have, they have their own, distinct, motivations as well. One such motivation (common to all these scholars irrespective of the field they work in, be it ecocriticism, environmental communication, anthrozoology, animal ethics, extinction studies or posthumanism) has to do with the fact that human thinking about animals constitutes a genuine psychological puzzle (Herzog 2010, 13). If this is so, then anything that can shape our thinking about other species (which is what narratives are alleged to be capable of doing), can also shed some light on its seemingly mysterious mechanisms, and as such is definitely worth scholarly attention.

But is this supposed mysteriousness of our thinking about animals being over-exaggerated? Not at all. In fact, a great deal of data has been amassed to support that claim. Consider the strange fact that, as the legal scholar Gary Francione constantly reiterates, contemporary Western society not only condemns billions of farm animals to a horrifying ordeal, but does so while at the same time professing the ideal of eradicating unnecessary suffering. Aren't we, officially, concerned with fighting cruelty and being humane to all living creatures (Francione 2000)? But isn't the suffering involved in meat production, for instance, utterly unnecessary? Francione thinks it is, just like "the 99,99999999%" of all the other kinds of suffering that the peace-loving citizens of the West condemn various other species to.[7] We are not sure about the exact number, but we definitely agree that the tension he describes is there (cf. Joy 2010).

If that tension is puzzling in itself, then it becomes even more so when one considers the fact that the Third Reich – the state which contemporary Western societies like to see as their historical antithesis and as the epitome of evil, inhumanity, and cruelty – had surprisingly progressive animal welfare legislation. The Nazi legal system was designed to reduce the suffering of species all across the board, from apes to crabs, while "the severity of the punishments" it mandated "was ... virtually unprecedented in modern times." As described by Arnold Arluke and Clinton Sanders in their book *Considering Animals*:

> On April 21, 1933, almost immediately after the Nazis came to power, the parliament passed a set of laws regulating the slaughter of animals. In August 1933, Hermann Göring announced an end to the "unbearable torture and suffering in animal experiments" and threatened to "commit to concentration camps those who still

think they can continue to treat animals as inanimate property"
.... He decried the "cruel" experiments of unfeeling scientists whose
allegedly unanesthetized animals were operated

...

In addition to the laws against vivisection and kosher slaughter,
other legal documents regulating the treatment of animals were en-
acted from 1933 through 1943, probably several times the number
promulgated in the previous half-century .... These documents covered in excruciating detail a vast array of concerns, from the shoeing
of horses to the use of anesthesia. One law passed in 1936 showed
"particular solicitude" about the suffering of lobsters and crabs.

(1996, 133–34)

How was it possible, ask Arluke and Sanders, that "the Nazis [could]
have been so concerned about cruelty to animals while they treated people so inhumanely" (132)? And how is it possible, let us add, that today,
in a country such as the USA, where 43 million citizens live in poverty
and where a similar number suffers from food insecurity, a few dozen
billion dollars is spent each year on pets, with a large part of this sum
used to buy luxury goods and services such as "massage therapy, ... New
Age animal communicators, ... Bowser Beer for dogs, [or] a 'Garden
Party Swarovski dress' for $3,000" (Herzog 2010, 75–76)? And how is
it possible that we care so much about some animals (not only pets, but
also certain undomesticated species such as the koala or panda), while
sending others "to slaughter with equanimity" (Rorty 1979, 190; cf.
DeMello 2012, 11)?

We could fill a large book with questions such as these, but, to reiterate, the point we want to make is that if human thinking about animals
is so puzzling in itself, then scholarly attention naturally turns to its
sources and mechanisms. In particular, it turns to how such thinking is
shaped and changed, which is exactly what animal narratives have been
reported to do — for better or for worse. If one wonders why certain irrational ways in which we treat animals seem so obviously reasonable to
so many people, then one is well advised to go back to all the childhood
stories which were in many cases our first source of information about
other species. Think of the stories about bad wolves, faithful dogs or, as
in Richard Scarry's bestselling children's books, about happy little pigs
who want to become butchers and who are fed bacon by their happy pig
mamas (Mansour 2005, 418; Gilson 2017). In the light of this, it is no
wonder that the increase in scholarly interest in human–animal relations
has been accompanied by the rise of studies on animal narratives all
across the academic spectrum, including in ecocriticism, literary animal
studies, extinction studies, animal ethics, and the like (see, e.g. Oswald
1995; Kerridge 2001; Huggan and Tiffin 2010; McHugh 2011; Gross
and Vallely 2012; Boggs 2013; Peterson 2013; Heise 2016; Barcz 2017;

Rose, Van Dooren, and Chrulew 2017; Weik von Mossner 2017; Barcz and Łagodzka, 2019).

Why Do We Want to Study Animal Stories?

While it should by now be clear why the impact of animal stories is an attractive subject for disciplines concerned with human–animal relations, there are also other fields in which it might be of considerable interest. In fact, while our own interest in it is mainly inspired by disciplines such as ecocriticism, animal studies, and extinction studies, it also stems from a field that is no less fascinating and at the same time much older than any of them. We are talking here about the study of the moral impact of stories (Gregory 2009).

What makes that field fascinating is that the subject of the moral influence of literature is situated at the juncture of fundamental questions about human morals, imagination, emotion, cognition, attitudes, and aesthetic preferences, the juncture at which many people find themselves at this or that point in their lives. We are certain, for instance, that each of the readers of this book must have wondered, at least once, why he or she reacts more emotionally to the fortunes and misfortunes of some fictional characters than to those of some real people and animals around her or him, or why people derive aesthetic pleasure from tales that depict the immense suffering of other people.

Some of our readers might have also wondered why there are always such heated debates about the shape of the literary canon whenever a significant ideological change occurs in society (Morrissey 2005; Bona and Maini 2006; Abate 2010). As we are writing these words, the students at Yale University, prompted by the recent waves of social justice movements in the USA, are demanding a reform of their literature curriculum, which they consider too white and heteronormative. Conversely in Poland, where the conservatives have recently taken power in government, there are complaints about certain books being in the school curriculum because they are supposedly too liberal, including *Harry Potter* (Flood 2016; Suchecka and Szpunar 2016)!

These and other questions about the moral impact of stories not only address something that is a common element of social life but also something deep about human minds. And it is precisely for this reason that the subject has been of considerable scholarly attention, most prominently in philosophy and literary studies. It is also not entirely surprising that it has been studied through the ages. As far as literary studies are concerned, we could draw a trajectory going at least from Horace's *Ars Poetica*, with its emphasis on the instructional role of poetic stories (Stock 2007, 94; cf. Habib 2011, chap. 3), to contemporary ethical criticism, one of whose most important representatives, Wayne Booth, argued that the moral influence that narratives exert upon us is best compared to that exerted by our friends (Booth 1988). In philosophy, there is an arc of thought which spans from Aristotle, who famously argued for the capacity of tragedy to provide us with moral illumination (Nussbaum 2001, 391), to the contemporary philosopher Richard Rorty, who equally famously argued that stories are great at making us think as "we" of "people whom we have previously thought of as 'they'" (1989, 192; cf. Johnson 2014). Or to put it more precisely, Rorty argued that narrative genres such as "the journalist's report, the comic book, the docudrama, and, especially, the novel" can help us to expand the circle of our moral concern:

> Fiction like that of Dickens, Olive Shreiner, or Richard Wright gives us details about kinds of suffering being endured by people to whom we had previously not attended. Fiction like that of Choderlos de Laclos, Henry James, or Nabokov gives us details about what sorts of cruelty we ourselves are capable of, and thereby lets us redescribe ourselves. That is why the novel, the movie, and the TV program have, gradually but steadily, replaced the sermon and the treatise as the principal vehicles of moral change and progress.
>
> (1989, xvi)

This alleged power of stories to expand the circle of moral concern has been on the mind not only of Rorty but also the minds of numerous other scholars, as well as writers and politicians. Their conviction that this power is real has even been translated into concrete educational policies. Consider the practice of assigning texts written by minority authors (Alice Walker's *The Color Purple* or Toni Morrison's *Beloved*) in school education or bibliotherapy as means of combating prejudices (Hinton and Dickinson 2007; Clark 2012; Montgomery and Maunders 2015).

Importantly, in doing so, educators can rely not only on their convictions but also on a large body of historical evidence. There are serious studies, most notably by the historian Lynn Hunt (2007) and the cognitive psychologist Steven Pinker (2011) that go even so far as to argue that one specific narrative form, the novel, has been a crucial factor behind the reduction or eradication of some forms of cruelty against disadvantaged groups which can be observed since the latter half of the eighteenth century, when Western societies witnessed an unprecedented expansion of the circle of empathic concern.

According to Hunt, before that time "[people] empathized with those close to them and with those most obviously like them – their immediate families, their relatives, the people of their parish, in general their customary equals." But then something happened and their empathy began to reach "across more broadly defined boundaries ... across class, sex, and national lines" (2007, 38). That something, she claims, had been intrinsically related to the rise of the novel and in particular to the enormous success of works such as Jean-Jacques Rousseau's *Julie* as well as Samuel Richardson's *Pamela* and *Clarissa*, which depicted those whom certain readers would typically exclude from the sphere of moral concern in situations that encouraged compassionate response (cf. Oatley 2011, 168–69).

Note that these novels not only exerted an emotional pressure that went against many readers' exclusivist intuitions, but simultaneously provoked them to lower their ideological guard, so to speak. In their desire to be entertained by a story, the reader was willing to suspend their prejudices against the social group represented by the main characters, and since an important part of being entertained by a novel is perspective taking, they would empathize with the protagonists, thereby becoming more susceptible to concern for other members of the group in question (cf. Strange 2002). To some, the experience was so shocking that it urged them to confide in the authors of those novels, something which Hunt's book reports in sometimes surprising detail:

> One Louis François, a retired military officer, wrote to Rousseau [about his reading of *Julie*]: "You have driven me crazy about her. Imagine then the tears that her death must have wrung from me. ... Never have I wept such delicious tears. The reading created such a powerful effect on me that I believe I would have gladly died during that supreme moment." Some readers explicitly acknowledged their identification with [Julie]. C.J. Panckoucke, who would become a well-known publisher, told Rousseau, "I have felt pass through my heart the purity of [her] emotions."
>
> (2007, 47–48)

From that time on, Hunt argues, it would be harder for the members of privileged classes to slight the suffering of peasants, servants, and members of various other disadvantaged groups, something which eventually led to the development of what we know today as human rights culture.

An analogous effect is sometimes attributed to another famous novel, Harriet Beecher Stowe's *Uncle Tom's Cabin* (1852), which is claimed to have significantly raised the concern of white Americans for the plight of their black compatriots (Morris, Sachsman, and Rushing 2007; Robbins 2007; Weinstein 2012). The novel was written as a response to a concrete and significant social fact – the introduction in 1850 of the Fugitive Slave Act, which "required all free people, North or South, to turn in escaped slaves directly to slave agents" (Ammons 2007, 7) – and it reportedly contributed to a number of other concrete and profound social facts too, including the military conflict between the North and South that left hundreds of thousands dead and eventually led to the liberation of millions. According to anecdote, when Abraham Lincoln first met Stowe, he greeted her with the words: "so you're the little lady who started the big war" (Ammons 2007, 187n1). If readers see parallels between the alleged wide social impact of *Uncle Tom's Cabin* and that exerted by the book with which we began, *Black Beauty*, then they are not alone. Sewell's novel has in fact been described as "the *Uncle Tom's Cabin* of the Horse," and it has encouraged the belief that literature can contribute to the reduction of the suffering inflicted by humans on non-human others (Nash 1989, 47).

It is here that the interests of scholars who study the ethics of stories and those who study the cultural aspects of human–animal and human–environment interactions potentially intersect. According to the American ecocritic Laurence Buell, one way in which literary stories have the capacity to extend our "environmental imagination" is by connecting us "vicariously with [the] experience, suffering, [and] pain ... of non-humans" (2001, 2). Another American ecocritic, Scott Slovic, adds that such an extension "might eventually have an impact on environmental laws and policies and on the daily behavior, even on the conscious and unconscious worldviews, of other members of society" (2008, 140).[8] In other words, these and other scholars believe that stories have a general capacity to do what *Black Beauty* did, to expand the circle of moral concern to include other species (cf. Singer 2011).[9]

That belief, for a number of reasons, is our interest in this book. First, from a theoretical point of view, there is something intriguing about the very idea of stories expanding the range of our moral concern *across* species lines, far more intriguing than the idea that they can expand that range *within* the confines of *Homo sapiens*. After all, we can quite easily understand how they do the latter. Consider, for instance, those examples of *Julie*, *Clarissa*, and *Uncle Tom's Cabin* mentioned above. The effect these stories achieved can be attributed in large part to their revealing to the reader from a privileged social group that members of given out-groups are in many morally relevant respects exactly like him or her, including in their mental capacities, something which was otherwise overshadowed by the superficial differences of skin color, education, profession, or wealth (Oatley 2011, 168–69). Following Lisa Zunshine, we might say that for such readers the novels worked as exercises in what the psychologists call *theory of mind* – "the ability to explain behavior in terms of the underlying states of mind" (2006, 4; cf. Bal and Veltkamp 2013; Rembowska-Płuciennik 2011). In other words, they improved their capacity to read the minds of others.

But note that among the many respects in which other species differ from us is that they cannot express their mental processes verbally, something which has historically encouraged many to doubt their morally relevant capacities, including the one to suffer (Serpell 1986, 160–61). How could readers believe that stories provide an insight into such capacities in animals? If stories can, despite this and other difficulties, raise our concern for other species, this would be truly remarkable.[10] Hence, our interest in the belief that they have this potential.

A second, more methodological reason for our interest is that this latter belief, however widespread, does not rest on much more than historical evidence, speculation, and anecdotal reports. In particular, it does not rest on statistically relevant empirical data, which is a serious drawback from a scholarly point of view. The main aim of this book is to remedy it by submitting the belief in question to empirical test.

But there is also a pragmatic motivation behind our project. It has to do with the fact that we agree with animal advocates and ethicists that the question of the unnecessary human-inflicted suffering of animals is of enormous importance today, given the massive scale on which such suffering is inflicted (Francione 2000; Singer and Mason 2006; Grimm 2015): There just *is* a pressing need to change that situation. And we also agree that the only way to do so involves changing the relevant attitudes of the public (Webster 2008; Corbey and Lanjouw 2013). If stories can help to achieve that, there are good reasons to think they might be of real importance for changing the current social realities of animal suffering. So there are good *practical* reasons to study their impact.

Why Is Empirical Evidence Important?

We have said above that the lack of statistically relevant evidence for the capacity of stories to impact our attitudes toward animals is a serious drawback. But why? Isn't it as clear as day that narratives *can* improve our attitudes toward other species? Who needs statistically relevant data to prove *that*? The best answer to such questions is that it is also as clear as day that the most commonly held intuitions about causal relations sometimes turn out to be wrong when submitted to empirical scrutiny, even if they had been supported by personal testimonies, historical data, and the like. Such intuitions include, e.g. the one that "ulcers are caused primarily or entirely by stress," or that when you turn 40 or 50, you will most likely "experience a midlife crisis" (Lilienfeld et al. 2010, 52–56, 126–29). Readers would be most surely surprised to learn how many of them have not been confirmed in rigorous studies, particularly those which concern human psychology.

Consider the commonly held belief that there is a causal link between animal cruelty in childhood and cruel behavior toward humans in adult life. It is often assumed that if one engages in cruel behavior toward animals as a child, then one will grow up more prone to cruelty toward humans than those who did not engage in such behavior in their childhood. As the US psychologist Harold Herzog notes, a belief in this causal link is "so well established that the term 'The Link' is now a registered trademark owned by the American Humane Association" (2010, 31).

However, he argues, the existing sociological data show no correlation between these two types of cruelty. For instance,

> Emily Patterson-Kane and Heather Pipe analyzed the results of two dozen research reports of childhood cruelty among extremely violent men ... and males with no history of violence... They found that 35% of the violent offenders had been childhood animal abusers – but so had 37% of the males in the 'normal' control group.
>
> (33)

And this is but one of the studies he mentions.

Cases such as The Link™ convey a very important message, and it is that we simply cannot rely on our intuitions and anecdotal data alone when it comes to general claims about causal relations in the social realm. We have to confront these claims with statistically relevant data. But importantly, while the kind of sociological data which Herzog used is sufficient to cast doubt on a causal claim, it would not be sufficient to prove it. For what Herzog was relying on are *correlational* studies, and such studies merely compare the occurrence of a given feature in a population with the occurrence of another one over time (Shaughnessy, Zechmeister, and Zechmeister 2012, 138). What transpired when he consulted such studies was that there was no correlation between childhood cruelty toward animals and adult cruelty toward humans. Therefore, he was entitled to have doubts about there being a causal link between these variables.

But what if those studies *had* shown a correlation between the variables in question? These would surely have been interesting results, but they would still not constitute sufficient proof that there is a cause-and-effect relation at play here. For instance, it might have been the case that people who engage in cruelty toward animals are simply more cruel in general – toward human beings and animals alike – so it is not that cruelty toward animals *caused* them to be more cruel toward humans. We could go on like this, but it should by now be clear that one problem with correlational studies is that there are always too many things we do not know about the studied subjects and their environment for the results to unanimously determine the causal mechanism behind the phenomenon we are focusing on. In other words, there are too many things which we cannot control, but should (Shaughnessy, Zechmeister, and Zechmeister 2012, 175–77).

How Do We Want to Study Stories?

The best-known method to avoid the general problem outlined above is the controlled experiment, where "controlled" means that the potential interfering factors are minimized (Webster and Sell 2007, 53–80).

How do such experiments allow us to confirm or disconfirm causal relations? Imagine our goal was to assess the value of taking vitamin C for fighting common cold. In order to conduct such an experiment we would have to, first, randomly assign patients suffering from the common cold to two groups: an experimental one, which takes vitamin C pills, and a control group, which is given a placebo, i.e. pills that do not contain any active substance. Importantly, neither group should know what exactly it is taking, which is why such studies are called "blind" (Cohen 2013, 199). (Ideally, in the case of "double-blind" studies, this knowledge is hidden also from the person administering the pills.) Finally, after the appropriate amount of time, both groups are submitted to tests and their results are compared. In this way, we can learn whether patients who take vitamin C get better sooner than those who did not take it.

The results of such an experiment can be seen as evidence for the causal impact of a drug or a lack thereof because of the high level of control they allow. Randomization precludes the possibility that the differences in results are due to differences between the groups (for example that people in one group were in a better condition than those in the other) and not due to the content of the pill that each group was administered (Cohen 2013, 200; Shaughnessy, Zechmeister, and Zechmeister 2012, 194–96). That people in the control group are given a placebo is crucial too. If the control group had not taken any pill, the differences in results might be explained away as perhaps due to the experimental group's expectations as to the power of the pill that they were administered rather than the active substance it contained (Shaughnessy, Zechmeister, and Zechmeister 2012, 200).

The blindness of the test, in turn, eliminates the possibility that the effects have been due to expectations on the part of the subjects as to the character of the active substance (Cohen 2013, 199–200). For instance, it has been shown that your belief that you have taken a stimulant may affect your blood-pressure levels (Harrington 1997). An analogous effect might occur if you believed you took something that is supposed to cure you of common cold, like vitamin C, for instance. If the test is *double*-blind, this eliminates the possibility that the placebo mechanism on the part of the patients is triggered by the behavior of the person administering the drug (Cohen 2013, 199). Such a person may unwittingly reveal to the patients through his or her words, gestures or other means what the expected results are (Shaughnessy, Zechmeister, and Zechmeister 2012, 201).

So this is how a properly controlled experiment on the influence of vitamin C on people suffering from the common cold would have to look in order for its results to be sound. And in case the readers wonder why anyone would even want to perform it (Isn't it as clear as day that vitamin C does help to fight common cold?), they should know that a study like this was conducted more than 20 years ago and it turned out that vitamin C does *not* help in this regard (Hemilä 1992). Here is further evidence that it is worth putting even our surest intuitions to experimental tests.

As for vitamin C, so for any other thing thought to have a causal impact on humans, including narratives and their impact on attitudes toward animals. So in order to study if this impact actually exists, we resorted to the experimental method, drawing from the field that specializes in producing experimental data on attitude change, that is, social psychology. We performed a series of more than a dozen controlled studies, involving over 3,000 subjects whom we submitted to literary narratives of various kinds: fictional, autobiographical, journalistic, and others. The subjects were always randomly assigned to an experimental group and a control group. The former read a narrative with animal-related content which we hypothesized might affect attitudes toward animals, while the control group read a narrative with content that was not related to animals and which we thought to be entirely neutral from the point of view of our study. Call it "narrative placebo." Our studies were also blind in that we did our best to prevent any possibility that the subjects realized what our purpose was, something which could have rendered our results unreliable.

That said, we do realize that reading a narrative is not entirely like taking a pill, and that one cannot study the former in exactly the same way in which one studies the latter. There just seem to be so many fleeting, subtle aspects about reception of stories that to study it experimentally seem to be akin to the experimental study of erotic intimacy or mystical rapture. Perhaps not necessarily a self-contradictory idea, but definitely a paradoxical one.

In order to face this and related problems, we have invested a lot of effort in constructing experimental designs that would possess what psychologists call a high level of ecological validity – in other words, studies that allow us to check if a given phenomenon occurs in real life or whether it can be observed only in artificial experimental conditions (Kellogg 2002, 120). This was made possible by our teaming up with a bestselling writer, Marek Krajewski, an author of dozens of detective novels translated into more than 20 languages, and a major literary star in our country, Poland.

Krajewski not only encouraged (actively and merely through his stardom) a large number of people to take part in our studies but also agreed to write an experimental narrative according to our suggestions! More than that, he included it in his novel – an actual novel which was then sold in bookstores, reviewed, discussed, and read by thousands. It was their reactions that we studied. For this and various other reasons, the results obtained in these experiments approximated in a remarkably accurate way the typical conditions in which people ordinarily read fiction. We like to see these studies as an unprecedented example of a happy marriage of literary art, the humanities, and social science, and we believe that the measures we had generally taken in conducting them ensure the soundness of their conclusions.

Why We, Too, Want to Tell a Story (and How We Are Going To Do It)

The present book will provide many concrete details about the data we gathered. But before that, a word is due on how it was written. First of all, we wanted this to be an academic monograph in the usual sense of presenting novel research and drawing on the current scholarly literature. And this is indeed what our book does. However, it also does something that academic monographs typically do *not* do. That is related to the fact that the novelty of our research is in large part of a methodological kind. We take certain instruments from one field (experimental social science) and apply it in another (the environmental humanities), where they have not been typically used.

One consequence of this is that we could not take for granted that most of our readers, whom we assumed would be environmental humanists, would be familiar with our methodology, the way authors of academic monographs, writing within and for a single discipline, normally can. So in order for our book to be useful for our audience we had to explain the *basics* of that methodology, often in a textbook fashion. Indeed, we would like to see this book as both an experimental contribution to the environmental humanities as well as a sort of introduction to how experimental methods can be profitably used in that field.

In writing the book, we also took into account the fact that these conclusions may be of interest not only to scholars, but also to readers who are not academics at all. These might include animal activists who want to know how to use stories in order to further their cause, or readers of those animal stories curious to know how they affect their attitudes and beliefs, or even readers interested in the psychological impact of stories in general. We therefore wanted to convey our scholarly message in a manner that would be understandable and interesting to laypeople.

This demanded using as little jargon as possible and explaining the meaning of technical vocabulary in all cases where we could not stay jargon-free. It also demanded explaining things which are trivial to readers who come from certain fields, but which may be unfamiliar to those who come from other fields, or from no academic field at all. So, in case any of our explanations seems annoying to the specialist reader in touching on things he or she feels everyone should know about, we would like to stress, again, that we included them in the book because we wanted it to be understandable also to non-specialists.

As researchers writing a book about narratives, it did not elude us also that our research, and any research for that matter, has its own narrative dimensions. As something that happens overtime, every experiment has its own chronology, and its description is always a story. Similarly, every research project has its own story too, and oftentimes such stories bear interesting similarities to the kind of stories Marek Krajewski specializes in (cf. Felski 2015, 85–116). After all, there is a deeper reason why the pursuit of truth by both detectives and researchers can be described by one English word – "investigation" (99).

So, similar to many criminal investigations, many scholarly investigations also have their own twists and turns, dead ends and sudden illuminations, moments that are suspenseful and moments that are revelatory. And just as many detectives will admit that, however exhausting such twists and turns may be, they are among the things they like best about their work, so will researchers. At any rate, this is part of what we enjoyed about our project, and to allow the reader to share at least part of that experience, we wrote the book as a narrative chronologically depicting the road we traveled.[11] This includes the moments when we seemed to be lost and those when we suddenly saw a new path opening. We decided to tell our story in this way also for the sake of scholarly transparency, so that everyone can see what was going on behind the scenes and better understand what follows from what. And finally, we would lie if we said that we did not think about that interesting article cited above that suggests that narrative style increases citations of scholarly works!

In Chapter 1, we will shed further light on the investigative techniques we used in order to confirm our hypotheses. A word will be said on the concept of attitudes and the instruments used to measure their change, on statistical analysis, and the difficulties pertaining to the experimental study of literature. Also, in this chapter, the reader will learn why all psychologists are liars, and why it is a very good thing that they are.

Chapter 2 will tell the story of our collaboration with Marek Krajewski and the resulting study that involved his readers, his publisher, his Facebook profile, and one of Poland's largest market research agencies. This study eventually received a lot of media attention when its results were first published (Małecki, Pawłowski, and Sorokowski 2016; cf. Burda 2017; Ślązak 2017).

Chapters 3–5 will present a series of laboratory experiments in which we studied the attitudinal impact of various internationally known animal stories, including Alice Walker's "Am I Blue?" and Marshall Saunders's *Beautiful Joe*. In all these laboratory experiments, in addition to testing our main hypothesis, we examined a number of secondary hypotheses, including whether there are differences in impact between stories which are perceived as fictional and those which are perceived as non-fictional, between stories written in first- and third-person voice, and whether there is a difference in impact between narratives whose animal protagonists belong to different species.

In order to explain the results of these studies, we will have to talk about a range of issues, including why narratives can make us think about others with parts of our brain that we normally use to think about ourselves, and why we should think of ourselves as elephants (and their riders). We will also talk about creatures as different as lizards, cows, and Pokémon, books as different as *Fifty Shades of Grey* and *Crime and Punishment*, and even about *Breaking Bad*.

But that it is not all. Chapter 5 will reveal how at some point we found ourselves in a moment of consternation that seems to be obligatory for any good detective story. The global data we gathered seemed to be inconclusive, and our investigation hung by a thread. There must have been a path we had somehow neglected, and Chapter 5 will explain how we found it.

In Chapter 6, we will explore the possible mechanisms behind the attitudinal impact of narratives, inquiring whether they might be related to engagement in the text and how cruel a story is. The readers will learn why being lost in a story can be deadly dangerous, and they will also be forced to admit that they enjoy reading about the suffering of other people and animals. But they will at the same be consoled that other people do that too.

Chapter 7 will turn to the question of how long animal stories can exert their influence on our minds. Is the impact an animal story makes upon you fleeting? Can it last a day? A week? Or perhaps even longer?

In the final part of the book, titled "Conclusions, Speculations, and Prospects," we will draw a number of scholarly and practical implications from our general results. As each investigation always opens new research horizons, we will also point to a few of those which were suggested by our project. And we will address a couple of worries that may be raised with regard to our research, including most importantly the very idea of using stories to change the way people think.

But one such potential worry should be addressed before we even begin as it might have occurred to some of our readers already. We mean the fact that, in order to confirm our hypotheses, we had to experiment on humans. It is a well-known fact that whenever experiments on humans are mentioned, most people become instinctively alarmed. They think about the potential for abuse, feel compelled to ask if the study is really necessary, and if they learn that it is, they then want to know whether all the necessary ethical standards were observed. And they are right to do so, given all the historical examples of clearly unethical or ethically dubious human experiments, including the infamous Tuskegee syphilis study and the Stanford prison experiment (Reverby 2000; Farrimond 2013, 1, 61–62). We would like to assure the readers, then, that we did our best to conduct our experiments with the utmost concern for our subjects, and that all the studies described in this book were approved by an appropriate Research Ethics Committee.

Finally, it is worth noting that while they were conducted on humans, the potential practical results of our experiments were meant to benefit mainly animals. In this way, our research admittedly strays from the usual scientific practice, where the reverse is the case. But this, as we are sure our readers will agree, is only fitting for a project investigating how our usual ways of thinking about animals might be changed.

Notes

1 Johnson and Johnson (2002, 254), cf. Davis (2016, 65–67) and Pearson (2011, 43–44).
2 See DeMello (2012, 183–84, 232), Garrard (2014, 409–22), Ortiz Robles (2016, 46–47), Dorré (2006, 95–120), Ramey (2011), Yeniyurt (2017), and Pearson (2011, 43–44, 124–25).
3 Needless to say, we do agree that humans belong to the animal kingdom, and that to refer collectively to representatives of other animal species as simply "animals" does not make any sense from a biological point of view. However, we choose to do so in this book for the simple reason that the consistent use of the most popular alternatives such as "non-human animals" and "other animals" would be too cumbersome. It should be noted here that there is at least one prominent thinker, Jacques Derrida, who believes that to use the term "animal" the way we do is "not simply a sin against rigorous thinking, vigilance, lucidity, or empirical authority, it is also a crime. Not a crime against animality, precisely, but a crime of the first order against the animals, against animals" (2009, 48). While we do see how such a use has been part and parcel of various ideologies which justify the oppression of animals, we allow ourselves to think that Derrida is slightly exaggerating here.
4 See also Grimm (2014, 2015).
5 See, e.g. the following publications, which address the narrative turn in fields as different as psychology, psychiatry, sociology, history, international relations, and political theory: Strange (2002), Lewis (2014), Berger and Quinney (2005), Cronon (1992), Roberts (2006), and Schiff (2014).
6 For a discussion of the phenomenon of people turning vegetarian allegedly as a result of having watched *Babe*, see Nobis (2009).
7 Admittedly, Francione gave this number in an e-mail correspondence with Mark Bekoff, cited in Bekoff (2010, 30).
8 One good historical example of the phenomenon Slovic talks about might be the social impact of US nature writing, see Philippon (2004).
9 It should be noted, however, that this belief is not universally accepted in the scholarly community. There are scholars who argue that it is not true, while some others remain skeptical about it arguing that it needs to be submitted to rigorous empirical tests before it can be considered sound. Perhaps, the most famous of the recent skeptics is Suzanne Keen, with the most famous of the recent expressions of the skeptical position being her book *Empathy and the Novel* (2007).
10 For an excellent theoretical discussion of narrative trans-species empathy, see chapter 4 of Weik von Mossner (2017).
11 In doing so, we may be seen as following the idea of narrative scholarship as defined by Scott Slovic. We definitely agree with him that "we must not reduce our scholarship to an arid and hyperintellectual game … devoid of actual experience," and that "We must analyze and explain literature through storytelling" (2008, 28).

Works Cited

Aaltola, Elisa. 2010. "Animal Ethics and the Argument from Absurdity." *Environmental Values* 19 (1): 79–98. doi:10.3197/096327110X485392.

Abate, Michelle Ann. 2010. *Raising Your Kids Right: Children's Literature and American Political Conservatism*. The Rutgers Series in Childhood Studies. New Brunswick, NJ: Rutgers University Press.

Actman, Jani. 2015. "12 Nat Geo Stories That Exposed Wildlife Exploitation." National Geographic. September 11, 2015. http://news.nationalgeographic. com/2015/11/151107-national-geographic-wildlife-crime-trafficking-animals/.

Ammons, Elizabeth. 2007. *Harriet Beecher Stowe's Uncle Tom's Cabin: A Casebook*. Oxford: Oxford University Press.

Arluke, Arnold, and Clinton Sanders. 1996. *Regarding Animals*. Philadelphia, PA: Temple University Press.

Bal, P. Matthijs, and Martijn Veltkamp. 2013. "How Does Fiction Reading Influence Empathy? An Experimental Investigation on the Role of Emotional Transportation." *PLOS ONE* 8 (1): e55341. doi:10.1371/journal. pone.0055341.

Barcz, Anna. 2017. *Animal Narratives and Culture: Vulnerable Realism*. Newcastle upon Tyne: Cambridge Scholars Publishing.

Barcz, Anna, and Dorota Łagodzka, eds. 2019. *Animals and Their People: Connecting East and West in Cultural Animal Studies*. Human-Animal Studies, vol. 21. Leiden; Boston: Brill.

Bekoff, Marc. 2010. *The Animal Manifesto: Six Reasons for Expanding Our Compassion Footprint*. Novato, CA: New World Library.

Berger, Ronald J., and Richard Quinney, eds. 2005. *Storytelling Sociology: Narrative as Social Inquiry*. Boulder, CO: Lynne Rienner Publishers.

Boggs, Colleen Glenney. 2013. *Animalia Americana: Animal Representations and Biopolitical Subjectivity*. Critical Perspectives on Animals: Theory, Culture, Science, and Law. New York: Columbia University Press.

Bona, Mary Jo, and Irma Maini, eds. 2006. *Multiethnic Literature and Canon Debates*. Albany: State University of New York Press.

Booth, Wayne C. 1988. *The Company We Keep: An Ethics of Fiction*. Berkeley: University of California Press.

Buell, Lawrence. 2001. *Writing for an Endangered World: Literature, Culture, and Environment in the U.S. and Beyond*. Cambridge, MA: Belknap Press of Harvard University Press.

Burda, Katarzyna. 2017. "Kryminał w Służbie Nauki. Treści Książek Zmieniają Sposób Myślenia." *Newsweek*, Accessed 2017. www.newsweek.pl/ plus/nauka/kryminal-w-sluzbie-nauki-tresci-ksiazek-zmieniaja-sposob-myslenia,artykuly,403956,1,z.html.

Camfield, Gregg. 2005. "The Sentimental and Domestic Traditions, 1865–1900." In *A Companion to American Fiction 1865–1914*, edited by Robert Paul Lamb and Gary Richard Thompson, 53–76. Malden, MA: Blackwell Publishing Ltd. doi:10.1002/9780470996829.ch4.

Clark, Anna. 2012. "How Toni Morrison's 'Beloved' Is Taught in Schools." *The Daily Beast*, Accessed October 4, 2012. www.thedailybeast.com/ articles/2012/10/04/how-toni-morrison-s-beloved-is-taught-in-schools.html.

Cohen, Barry H. 2013. *Explaining Psychological Statistics*, 4th ed. Coursesmart. Hoboken, NJ: Wiley.

Corbey, Raymond, and Annette Lanjouw, eds. 2013. *The Politics of Species: Reshaping Our Relationships with Other Animals*. Cambridge: Cambridge University Press.

Cosslett, Tess. 2006. *Talking Animals in British Children's Fiction, 1786–1914*. The Nineteenth Century Series. Aldershot; Burlington, VT: Ashgate.

Cronon, William. 1992. "A Place for Stories: Nature, History, and Narrative." *The Journal of American History* 78 (4): 1347–76. doi:10.2307/2079346.

Davis, Janet M. 2016. *The Gospel of Kindness: Animal Welfare and the Making of Modern America*. Oxford; New York: Oxford University Press.

Dawkins, Marian Stamp. 2008. "The Science of Animal Suffering." *Ethology* 114 (10): 937–45. doi:10.1111/j.1439-0310.2008.01557.x.

DeMello, Margo. 2012. *Animals and Society: An Introduction to Human-Animal Studies*. New York: Columbia University Press.

Derrida, Jacques. 2009. *The Animal That Therefore I Am*. Translated by David Wills. Fordham: Fordham University Press.

Dorré, Gina M. 2006. *Victorian Fiction and the Cult of the Horse*. Aldershot; Burlington, VT: Ashgate.

Eitler, Pascal. 2014. "Doctor Dolittle's Empathy." In *Learning How to Feel: Children's Literature and Emotional Socialization, 1870–1970*, edited by Ute Frevert, Pascal Eitler, and Stephanie Olsen, 1st ed., 94–114. Emotions in History. Oxford: Oxford University Press.

Elżanowski, Andrzej. 2013. "Moral Progress: A Present-Day Perspective on the Leading Enlightenment Idea." *Argument* 3 (1): 9–26.

Farrimond, Hannah. 2013. *Doing Ethical Research*. Houndmills, Basingstoke, Hampshire; New York: Palgrave Macmillan.

Felski, Rita. 2015. *The Limits of Critique*. Chicago: The University of Chicago Press.

Flood, Alison. 2016. "Yale English Students Call for End of Focus on White Male Writers." *The Guardian*, Accessed June 1, 2016, sec. Books. www.theguardian.com/books/2016/jun/01/yale-english-students-call-for-end-of-focus-on-white-male-writers.

Foer, Jonathan Safran. 2009. *Eating Animals*. 1st ed. New York: Little, Brown and Company.

Francione, Gary L. 2000. *Introduction to Animal Rights: Your Child or the Dog?* Philadelphia, PA: Temple University Press.

Garrard, Greg, ed. 2014. *The Oxford Handbook of Ecocriticism*. New York, NY: Oxford University Press.

Gilson, David. 2017. "This Little Piggy Had Roast Beef: Animal Cannibalism in the Beloved World of Richard Scarry." *Davegilson.Com*, 2017. http://davegilson.com/scarry.html.

Greenfield, Nicole. 2016. "Poachers Beware: Cambodia's Endangered Wildlife Has a Passionate Defender." *NRDC*, Accessed July 21, 2016. www.nrdc.org/stories/poachers-beware-cambodias-endangered-wildlife-has-passionate-defender.

Gregory, Neville G. 2008. *Physiology and Behaviour of Animal Suffering*. New York: John Wiley & Sons.

Gregory, Marshall W. 2009. *Shaped by Stories: The Ethical Power of Narratives*. Notre Dame, IN: University of Notre Dame Press.

20 *Introduction*

Grimm, David. 2014. "Animal Welfare Accreditation Called into Question." *Science* 345 (6200): 988. doi:10.1126/science.345.6200.988.

———. 2015. "The Insurgent." *Science* 347 (6220): 366–69. doi:10.1126/science.347.6220.366.

Gross, Aaron, and Anne Vallely, eds. 2012. *Animals and the Human Imagination: A Companion to Animal Studies.* New York: Columbia University Press.

Habib, Rafey. 2011. *Literary Criticism from Plato to the Present: An Introduction.* Malden, MA: John Wiley & Sons.

Haidt, J. 2001. "The Emotional Dog and Its Rational Tail: A Social Intuitionist Approach to Moral Judgment." *Psychological Review* 108 (4): 814–34.

Harrington, Anne, ed. 1997. *The Placebo Effect: An Interdisciplinary Exploration.* Cambridge, MA: Harvard University Press.

Heise, Ursula K. 2016. *Imagining Extinction: The Cultural Meanings of Endangered Species.* Chicago, IL; London: The University of Chicago Press.

Hemilä, Harri. 1992. "Vitamin C and the Common Cold." *British Journal of Nutrition* 67 (1): 3–16. doi:10.1079/BJN19920004.

Herzog, Harold. 2010. *Some We Love, Some We Hate, Some We Eat: Why It's so Hard to Think Straight about Animals.* 1st ed. New York, NY: Harper.

Hillier, Ann, Ryan P. Kelly, and Terrie Klinger. 2016. "Narrative Style Influences Citation Frequency in Climate Change Science." *PLOS ONE* 11 (12): e0167983. doi:10.1371/journal.pone.0167983.

Hinton, KaaVonia, and Gail K Dickinson. 2007. *Integrating Multicultural Literature in Libraries and Classrooms in Secondary Schools.* Columbus, OH: Linworth Pub.

Huggan, Graham, and Helen Tiffin. 2010. *Postcolonial Ecocriticism: Literature, Animals, Environment.* London; New York: Routledge.

Hunt, Lynn. 2007. *Inventing Human Rights: A History.* 1st ed. New York: W.W. Norton & Co.

"Interview with Gail Eisnitz, Author of 'Slaughterhouse.'" n.d. A Vegan Skeptic. Accessed January 26, 2017. www.wegodlessanimals.com/inteview-with-gail-eisnitz-author-of-slaughterhouse/.

Johnson, Claudia Durst, and Vernon E. Johnson. 2002. *The Social Impact of the Novel: A Reference Guide.* Westport, CT: Greenwood Press.

Johnson, Peter. 2014. *Moral Philosophers and the Novel: A Study of Winch, Nussbaum and Rorty.* Houndmills, Basingstoke, Hampshire; New York: Palgrave Macmillan.

Joy, Melanie. 2010. *Why We Love Dogs, Eat Pigs, and Wear Cows: An Introduction to Carnism: The Belief System That Enables Us to Eat Some Animals and Not Others.* San Francisco, CA: Conari Press.

Keen, Suzanne. 2007. *Empathy and the Novel.* Oxford; New York: Oxford University Press.

Kellogg, Ronald T. 2002. *Cognitive Psychology.* London: SAGE Publications.

Kerridge, Richard. 2001. "Ecological Hardy." In *Beyond Nature Writing: Expanding the Boundaries of Ecocriticism*, edited by Karla Armbruster and Kathleen R. Wallace, 126–42. Under the Sign of Nature. Charlottesville: University Press of Virginia.

Kreuter, Matthew W., Melanie C. Green, Joseph N. Cappella, Michael D. Slater, Meg E. Wise, Doug Storey, Eddie M. Clark, et al. 2007. "Narrative

Communication in Cancer Prevention and Control: A Framework to Guide Research and Application." *Annals of Behavioral Medicine: A Publication of the Society of Behavioral Medicine* 33 (3): 221–35. doi:10.1080/08836610701357922.

Lewis, Bradley. 2014. "Taking a Narrative Turn in Psychiatry." *The Lancet* 383 (9911): 22–23. doi:10.1016/S0140-6736(13)62722-1.

Lilienfeld, Scott O., Steven Jay Lynn, John Ruscio, and Barry L. Beyersteub. 2010. *50 Great Myths of Popular Psychology: Shattering Widespread Misconceptions about Human Behavior*. Chichester, West Sussex; Malden, MA: Wiley-Blackwell.

Małecki, Wojciech, Bogusław Pawłowski, and Piotr Sorokowski. 2016. "Literary Fiction Influences Attitudes Toward Animal Welfare." *PLOS ONE* 11 (12): e0168695. doi:10.1371/journal.pone.0168695.

Mansour, David. 2005. *From Abba to Zoom: A Pop Culture Encyclopedia of the Late 20th Century*. Kansas City, MO: Andrews McMeel Publishing.

Masson, Jeffrey Moussaieff. 2008. *The Pig Who Sang to the Moon: The Emotional World of Farm Animals*. New York: Random House.

McGinn, Colin. 1997. *Minds and Bodies: Philosophers and Their Ideas*. New York: Oxford University Press.

McHugh, Susan. 2011. *Animal Stories: Narrating across Species Lines*. Posthumanities, v. 15. Minneapolis: University of Minnesota Press.

Montgomery, Paul, and Kathryn Maunders. 2015. "The Effectiveness of Creative Bibliotherapy for Internalizing, Externalizing, and Prosocial Behaviors in Children: A Systematic Review." *Children and Youth Services Review* 55 (August): 37–47. doi:10.1016/j.childyouth.2015.05.010.

Morris, Roy, David B Sachsman, and S. Kittrell Rushing. 2007. *Memory and Myth: The Civil War in Fiction and Film from Uncle Tom's Cabin to Cold Mountain*. West Lafayette, IN: Purdue University Press.

Morrissey, Lee, ed. 2005. *Debating the Canon: A Reader from Addison to Nafisi*. 1st ed. New York: Palgrave Macmillan.

Nash, Roderick. 1989. *The Rights of Nature: A History of Environmental Ethics*. History of American Thought and Culture. Madison: University of Wisconsin Press.

Nobis, Nathan. 2009. "The 'Babe' Vegetarians: Bioethics, Animal Minds and Moral Methodology." In *Bioethics at the Movies*, edited by Sandra Shapsay, 56–73. Baltimore, MD: Johns Hopkins University Press.

Nussbaum, Martha Craven. 2001. *The Fragility of Goodness: Luck and Ethics in Greek Tragedy and Philosophy*. Cambridge: Cambridge University Press.

Oatley, Keith. 2011. *Such Stuff as Dreams: The Psychology of Fiction*. Chichester, West Sussex, UK; Malden, MA: Wiley-Blackwell.

O'Connor, Amy. 1995. "When Pigs Fly." *Vegetarian Times*, December, 16.

O'Connor, Ellen. 2002. "Storied Business: Typology, Intertextuality, and Traffic in Entrepreneurial Narrative." *The Journal of Business Communication (1973)* 39 (1): 36–54. doi:10.1177/002194360203900103.

Ortiz Robles, Mario. 2016. *Literature and Animal Studies*. New York: Routledge.

Oswald, Lori Jo. 1995. "Heroes and Victims: The Stereotyping of Animal Characters in Children's Realistic Animal Fiction." *Children's Literature in Education* 26 (2): 135–49.

Pearson, Susan J. 2011. *The Rights of the Defenseless: Protecting Animals and Children in Gilded Age America*. Chicago, IL: University of Chicago Press.

Peterson, Christopher. 2013. *Bestial Traces: Race, Sexuality, Animality*. 1st ed. New York: Fordham University Press.

Philippon, Daniel J. 2004. *Conserving Words: How American Nature Writers Shaped the Environmental Movement*. Athens: University of Georgia Press.

Pijoos, Iavan. 2016. "White Rhino Survives North West Poaching." *News24*, Accessed September 26, 2016. www.news24.com/SouthAfrica/News/white-rhino-survives-north-west-poaching-20160926.

Pinker, Steven. 2011. *The Better Angels of Our Nature: Why Violence Has Declined*. New York: Viking.

Pollock, Mary Sanders. 2005. "Ouida's Rhetoric of Empathy: A Case Study in Victorian Anti-Vivisection Narrative." In *Figuring Animals: Essays on Animal Images in Art, Literature, Philosophy and Popular Culture*, 135–59. Palgrave Macmillan, New York. doi:10.1007/978-1-137-09411-7_9.

Ramey, David. 2011. "A Historical Survey of Human–Equine Interactions." In *Equine Welfare*, edited by C. Wayne McIlwraith and Bernard E Rollin. Oxford: Wiley-Blackwell.

Rembowska-Płuciennik, Magdalena. 2011. "Narrative Poetics of Mindreading." *Zagadnienia Rodzajów Literackich* 54 (1): 175–90.

Reverby, Susan, ed. 2000. *Tuskegee's Truths: Rethinking the Tuskegee Syphilis Study*. Studies in Social Medicine. Chapel Hill, NC: University of North Carolina Press.

Robbins, Sarah. 2007. *The Cambridge Introduction to Harriet Beecher Stowe*. Cambridge; New York: Cambridge University Press. www.myilibrary.com?id=81556.

Roberts, Geoffrey. 2006. "History, Theory and the Narrative Turn in IR." *Review of International Studies* 32 (4): 703–14. doi:10.1017/S0260210506007248.

Rorty, Richard. 1979. *Philosophy and the Mirror of Nature*. Princeton: Princeton University Press.

———. 1989. *Contingency, Irony, and Solidarity*. Cambridge ; New York: Cambridge University Press.

———. 1998. *Truth and Progress*. Philosophical Papers, v. 3. Cambridge ; New York: Cambridge University Press.

Rose, Deborah Bird, Thom Van Dooren, and Matthew Chrulew, eds. 2017. *Extinction Studies: Stories of Time, Death, and Generations*. New York: Columbia University Press.

Salmon, Christian. 2010. *Storytelling: Bewitching the Modern Mind*. London; Brooklyn, NY: Verso.

"Saving the Survivors." 2017. 2017. www.savingthesurvivors.org/.

Savvides, Nikki. 2013. "Speaking for Dogs: The Role of Dog Biographies in Improving Canine Welfare in Bangkok, Thailand." In *Speaking for Animals: Animal Autobiographical Writing*, edited by Margo DeMello, 1st ed., 232–43. Routledge Advances in Sociology 80. New York, NY: Routledge.

Schiff, Jade. 2014. *Burdens of Political Responsibility: Narrative and the Cultivation of Responsiveness*. New York, NY: Cambridge University Press.

Serpell, James. 1986. *In the Company of Animals: A Study of Human-Animal Relationships*. Oxford [Oxfordshire]; New York, NY, USA: B. Blackwell.

Sewell, Anna. 2012. *Black Beauty*. Oxford: Oxford University Press.

Shaughnessy, John J, Eugene B Zechmeister, and Jeanne S Zechmeister. 2012. *Research Methods in Psychology*. New York, NY: McGraw-Hill.

Singer, Peter. 2009. *Animal Liberation: The Definitive Classic of the Animal Movement*. Updated ed., 1st Ecco pbk. ed., 1st Harper Perennial ed. New York: Ecco Book/Harper Perennial.

———. 2011. *The Expanding Circle: Ethics, Evolution, and Moral Progress*. 1st Princeton University Press pbk. ed. Princeton, NJ: Princeton University Press.

Singer, Peter, and Jim Mason. 2006. *The Ethics of What We Eat: Why Our Food Choices Matter*. Emmaus, PA; New York: Rodale.

Ślązak, Anna. 2017. "To Nie Truizm. Czytanie Naprawdę Zmienia Świat." Nauka w Polsce. January 12, 2017.

Slovic, Scott. 2008. *Going Away to Think: Engagement, Retreat, and Ecocritical Responsibility*. Reno: University of Nevada Press.

Stock, Brian. 2007. *Ethics Through Literature: Ascetic and Aesthetic Reading in Western Culture*. Hanover: University Press of New England.

Strange, Jeffrey J. 2002. "How Fictional Tales Wag Real-World Beliefs." In *Narrative Impact: Social and Cognitive Foundations*, edited by Melanie C. Green, Jeffrey J. Strange, and Timothy C. Brock, 262–86. Mahwah, NJ: L. Erlbaum Associates.

Suchecka, Justyna, and Olga Szpunar. 2016. "Czym Zgrzeszył Harry Potter. Lektury Szkolne Na Indeksie - Rozmowa z Dr. Hab. Andrzejem Waśką." *Gazeta Wyborcza*, October 12, 2016. http://wyborcza.pl/magazyn/7,124059,21098142,czym-zgrzeszyl-harry-potter-lektury-szkolne-na-indeksie.html.

Webster, John. 2008. *Animal Welfare: Limping Towards Eden*. New York: John Wiley & Sons.

Webster, Murray, and Jane Sell, eds. 2007. *Laboratory Experiments in the Social Sciences*. Amsterdam; Boston, MA: Academic Press/Elsevier.

Weik von Mossner, Alexa. 2017. *Affective Ecologies: Empathy, Emotion, and Environmental Narrative*. Cognitive Approaches to Culture. Columbus: The Ohio State University Press.

Weinstein, Cindy. 2012. *Cambridge Companion to Harriet Beecher Stowe*. Cambridge: Cambridge University Press.

Yeniyurt, Kathryn. 2017. "Black Beauty: The Emotional Work of Pretend Play." In *Animals in Victorian Literature and Culture*, 233–49. Palgrave Studies in Animals and Literature. London: Palgrave Macmillan. doi:10.1057/978-1-137-60219-0_12.

Zunshine, Lisa. 2006. *Why We Read Fiction: Theory of Mind and the Novel*. Columbus: Ohio State University Press.

1 Texts, Statistics, and Deception

On Our Investigative Method

How Not to Chase a Chimera

One of the foundational texts of detective fiction, and a classic that continues to be revered, is Edgar Allan Poe's "The Purloined Letter." The plot of this short story is deceptively simple. An epistle is stolen from the French Queen containing information that could be extremely harmful to her if revealed to the third parties. The Police know that the perpetrator is Minister D., who uses the letter to blackmail the Queen, and that he keeps it in the hotel apartment where he lives. For various reasons, they cannot arrest or even interrogate him, and the only available course of action is to try to retrieve the letter secretly from that hotel. So, whenever the minister leaves the hotel, which he does frequently, a team of police agents meticulously scours the apartment and the rest of the hotel.

This team of agents searches all of the places it seems the letter could be hidden. They examine "the furniture of each apartment" and open "every possible drawer." They "scrutinize each individual square inch throughout the [hotel] including the two houses immediately adjoining, with the microscope." They examine "the moss between the bricks," open "every package and parcel" and "every book." They remove "every carpet," examine the cellars, look to "the mirrors, between the boards and the plates," and "the bed-clothes". They look behind "the curtains and carpets," too (Poe 1994, 342–44).

This work continues for months but without success. And just as the inspector in charge begins to despair (and the reader with him), the solution is provided by an amateur detective, C. Auguste Dupin, in what has become one of the most famous unveilings in all detective fiction. We learn that the entire time the police had been searching the hotel, the epistle was there, right before their eyes, placed in a card-rack "that hung dangling by a dirty blue ribbon" in one of the rooms that had been examined so methodically (Poe 1994, 353). In other words, it was *not* hidden at all, at least not in the sense which the policemen assumed.

Like any other classic, "The Purloined Letter" can be profitably read in a number of ways, and this potential has been exploited by numerous commentators, including such prominent figures as Jacques Lacan,

DOI: 10.4324/9780429061424-2

Jacques Derrida, and Barbara Johnson (Muller and Richardson 1987; Felski 2015, 85). But the message we find most interesting given the purpose of our book is the deceptively simple one that if you look for something, then whether you are a detective or a scholar or anybody else, you have to begin by clearly explaining to yourself what it is exactly that you are looking for. Unless you do, then no matter how methodically you search, you may end up empty-handed.

The policemen in the story were apparently looking for a letter belonging to the Queen, but they also assumed (quite naturally) that they were looking for something *hidden*. This is why they literally could not recognize the letter that was placed in plain sight as the one they were there for. This is also why they owe the failure of their search to themselves: they were looking for a hidden letter, but no such letter existed. They were after a figment of their own imagination, a ghost, a chimera. All their critical intelligence, skills, equipment, and experience could not help them, and all because of the simple mistake they made in the very beginning.

The continuing relevance of Poe's story lies in part in the universal character of the mechanism it describes. For some reason, humans have always been prone to make the kind of mistake the police detectives committed: not only in criminal investigations, but in virtually any area of life, including science. The history of any scholarly field is replete with stories of huge projects collapsing precisely because researchers uncritically assumed something about their subject that turned out to be entirely wrong, or that their assumptions about it were too vague to lead to any concrete results (Becker 2014; Loeb 2014). This is why we called the methodological message of the short story *deceptively* simple. None of us thinks we need to be reminded of such a triviality ("Of course I know what I am looking for!"), but somehow we forget about it every single day. This is why practically all Psychology 101 students are dutifully reminded by their professors to first precisely define the phenomenon they want to study before they embark on gathering data. And this is why professors need to remind themselves about it too.

In the Introduction, we used a couple of different formulations in order to describe the subject of our study. We said that we are interested in whether stories can raise our concern for the well-being of animals, shape the way we think about it, and can have an impact on our attitudes toward it. All these phrases seem to refer to one phenomenon, but the contours thereof seem to be blurred. As we did not want to end up chasing a chimera, we began our project by defining what it is that we are after, and we eventually agreed that we should focus on measuring the impact of literary stories on attitudes toward the well-being of animals.

Admittedly, such an influence is not everything that scholars and animal advocates refer to when they talk about the moral or social impact of animal stories. They also talk about plenty of other things such as an

increase in empathy, the expansion of imagination, the raising of awareness, the changing of behavior, and the like. We agree with them that all
these things are real phenomena, and that our focus merely on attitudes
will present only a partial picture of what the moral impact of animal
stories consists in. But still, that part of the picture is important, and
moreover, it can be defined precisely enough to yield concrete results
when studied.

The main reason it can be defined with such a precision is that attitudes belong to one of the most thoroughly researched social psychological phenomena. In fact, there was a time, back in the 1920s, when "such
was the importance of work on attitude measurement that that social
psychology was often defined as the study of attitudes" (Maio and Haddock 2012, 5). While it cannot accurately be defined in this way now,
there is no doubt that as far as attitudes are concerned we can take advantage of a massive body of work produced over more than a century of
research (Eagly and Chaiken 1993; Eagly and Chaiken 1998; Crano and
Prislin 2008). What this means is that researchers have had enough time
to produce sound conceptualizations of attitudes and develop sophisticated techniques for measuring them, not to mention commit numerous
errors from which we can learn today.

Attending to Attitudes

The definition of "attitude" that we assume in this work is the standard
psychological one, which states that an attitude is "a psychological tendency that is expressed by evaluating a given entity with some degree
of favor or disfavor" (Eagly and Chaiken 1993, 1). If an individual is
prone to conceive of X as possessing a given value (positive, negative,
or anything in between), then he or she can be said to have an attitude
toward X. All of us harbor attitudes toward numerous things – be they
tomato soup, rats, or the philosophy of Kant – and attitudes constitute a
fundamental part of our identity. Just think about how often we define
ourselves in terms of whether we value this or that, and how often we
judge others on the basis of whether they attach an equal value to those
things. Why is it you think detective novels are trash? How can Andrew
enjoy hunting? Why is Beverly in favor of euthanasia? How can anyone
savor *foie gras*? If you consider this, and if you additionally consider how
much social conflict is due to differences in attitudes and the great extent
to which social cooperation is dependent on managing attitudes, then it
becomes clear why social psychology is so concerned with attitudinal
change and with how to measure it.

Just as in our project we adopt a textbook understanding of attitudes,
we also adopt a textbook, tried-and-tested, way to measure them. The
method relies on so-called scales: sets of questionnaire items which allow only for fixed, quantifiable responses, where the total score achieved

on all the items indicates the extent to which a person harbors a given attitude (Maio and Haddock 2012, 10–12). A typical item in a scale would be a statement of a particular belief, accompanied by a range of answers describing different levels of agreement with that statement. For instance, the statement might be something like "The slaughter of whales and dolphins should be immediately stopped even if it means that some people will be put out of work," plus a selection of answers that indicate how strongly or otherwise one believes the statement, ranging from "Strongly agree," through "Agree," "Undecided," and "Disagree" to "Strongly Disagree," where these answers are ascribed points from five to one, respectively. That particular example is in fact taken from an actual scale which was developed by Harold Herzog and colleagues in order to capture whether one's attitude is more or less sympathetic toward animal welfare (Herzog, Betchart, and Pittman 1991). The higher the score achieved on all the items in that scale (called the Animal Attitudes Scale, or AAS), the more pro-animal one can be taken to be. There are 20 items altogether in this scale which means that the total score can range anything from 20 (the least pro-animal welfare) to 100 (the most pro-animal).

But note that not all items in the AAS are like the one about dolphins and whales, where a higher score obviously represents a more pro-animal attitude than a lower one would. Consider, for instance, the following item: "Basically, humans have the right to use animals as we see fit." Obviously, if one strongly agreed with this one, and scored more points on it, then we would be entitled to suspect that the person in question is *less* pro-animal welfare than a person who strongly disagreed with it. Such items are "reverse scored" because in calculating the summary score achieved on the scale, one has to translate the score achieved on these items by reversing it. For instance, if somebody replied "Strongly disagree" to the item on using animals as we see fit, then they would receive five points, not one; if they replied "Disagree," they would receive four, not two, points, and so on.

Such items are to be found not only in Herzog's scale, but in practically all scales used by psychologists. And if one wonders why they are there, then the answer is not that psychologists like to make life difficult for themselves, but rather that their subjects often make life difficult for them. It so happens, *terribile dictu*, that not all people who volunteer to participate in psychological studies value the scientific truth enough to fill the questionnaires the way they are supposed to be filled, that is, by reading carefully all the items and giving thought out and sincere answers to them. Some do not feel like making that kind of effort and answer randomly or by applying a kind of system that allows them to answer the questions automatically by giving the same, or roughly the same answer to each item. Moreover, one cannot rule out that among one's participants there will be people incapable for this or that reason of fully grasping the meaning of the items. By putting reversed items in their questionnaires, psychologists are able to detect such respondents and reject their questionnaires as invalid (Maio and Haddock 2012, 12). For instance, if somebody replied "Strongly agree" to all the items on Herzog's scale (including the one on dolphins *and* the one on using animals as we see fit), then that would mean that he or she was either cheating or did not understand the items. Either way, no researcher should take such a questionnaire into account.

But in order to be trustworthy, scales need not only to contain reversed items but, most importantly, have to meet the criteria of validity and internal consistency. "Internal consistency" concerns the correlation between replies to particular items within the scale. In other words, if a scale measures the same phenomenon, (or construct, as psychologists like to say), then the scores for particular items should correlate when tested on a larger group of people (Maio and Haddock 2012, 20–22). For instance, the subjects' scores on reversed items should be lower than those on the non-reversed items, and they also should be similar within each category. If they are not, then we are entitled to judge that the scale which includes them is inconsistent and therefore useless as a psychometric tool.

Validity, in turn, means that a scale picks out exclusively, and at the same time comprehensively, the phenomenon it aims to study (Maio and Haddock 2012, 20–22). Thus understood, validity may seem to be so obvious a criterion that it does not even merit stating, but the fact that in practice it is very often violated (including by experienced researchers) indicates that the question does demand special attention. As to how easy it is to err in this regard, consider the hypothetical item, "I enjoy watching wild animals in their natural habitat," which might prima facie seem a good candidate for inclusion in a scale measuring attitudes toward animal welfare (at least insofar as it apparently constitutes the opposite of the item "I sometimes feel upset when I see animals in cages at zoos," which does appear in the AAS). However, "enjoy" is an ambiguous term that denotes all kinds enjoyment (moral, aesthetic, cognitive, etc.), and for this reason we cannot be sure that the respondents will understand our item in a way that could reveal the level of their concern for animal welfare.

For instance, if "I enjoy watching wild animals in their natural habitat" is understood in purely aesthetic terms, it will have nothing to do with concern for the welfare of these animals because taking pleasure in the aesthetic qualities possessed by a living being does not *necessitate* caring for its welfare. And conversely, if someone "enjoys" watching wild animals in their natural habitat in the sense that he or she is happy that these animals can fully exercise their natural capacities, then this may have nothing to do with appreciating aesthetically what he or she sees. Thus, both a committed animal activist and a recreational hunter may reply to that particular question in the very same way, which may negatively affect the validity of the scale as a whole. One lesson that flows from this example is to avoid ambiguous statements since their content may be filled out by the participants in many different ways.

The AAS, which we have been referencing throughout this sub-chapter, does a fine job at avoiding ambiguity, and it is also possessed of a very high level of internal consistency (H. A. Herzog and Mathews 1997, 171). This is one of the reasons why it has been used by various scholars across different fields (Flynn 2003; Rothgerber 2014; Lee et al. 2015). However, despite its being a prima facie natural choice for our project,[1] we could not use it for two reasons. The first one is its cultural specificity. It was developed in the USA and bears the cultural marks of that country to such an extent that some of its items might be understood differently from how Herzog and colleagues intended in the specifically Polish cultural context in which we worked. Some statements in the AAS concern rodeos and cock-fighting, which are not popular in Poland, and some mention animals like raccoons, which are not widely known here.

Another reason is that even in its shortened version, which includes 20 items, it was simply too long for our purposes. Not too long because, say, the subjects would become tired with answering the items, which is indeed the problem with some questionnaires, but too long for us to hide the scale from them.[2] Yes, we have to admit that we were out to hide something from our subjects. But explaining why we wanted to deceive them and why we had to design our own scale to do so demands a separate discussion.

Why Psychologists Are Such Liars (and Why It's a Good Thing)

One of the things for which the German philosopher Immanuel Kant was famous was the uncompromising prohibition on lying and deception that he advocated. And we mean here *literally* uncompromising in the sense that allows for *no exceptions* whatsoever: according to Kant, you should never lie or deceive, regardless of the circumstances. As he sternly argued (1993 [1797]), even if a murderer armed with an axe was chasing your friend and asked you about his whereabouts, you would be duty-bound to tell the murderer the truth, contributing to your friend's demise, and forbidden from lying in order to protect them.

Needless to say, if people followed Kant in this regard, the world would look very different. And one of the less immediately obvious ways in which it would differ is that in such a world social psychology as we know it could not exist – to such an extent it relies on deception, including straightforward lying (Shaughnessy, Zechmeister, and Zechmeister 2012, 72–76). The reason it seems to be unabashed about this (and at the same time the reason why we are not afraid to say that sometimes deception is good) is that in psychology, lying is used mainly for the pursuit of truth, the truth that is both important and which cannot be achieved otherwise. The latter is because of the simple fact that in psychology, unlike in, say, physics or chemistry, the subject of your study is usually capable of understanding the reasons for your inquiry, and this understanding may influence the results. So, this subject must be kept ignorant of your aims and what is going on. They will have to be deceived.

Consider studying attitudes. If a subject of an experiment realizes what the study is concerned with, he or she may behave in a way that will reflect more the way she wants to be perceived by researchers than her actual dispositions. This is sometimes called the problem of "impression management" (Maio and Haddock 2012, 14; Tedeschi 2013) and is especially likely to occur in studies like ours, measuring a predisposition which is positively evaluated in the culture to which the subjects belong. After all, everybody knows that being good to animals is a good thing. And of course, nobody wants to be perceived as cruel, callous, or merely indifferent.

This mechanism, we may hypothesize, seems also to be the main factor behind the moral schizophrenia which Gary Francione (2000) talks about, where everybody *declares* kindness to animals, yet allows for monstrous unkindness to take place. Though Francione has been rightly criticized for his use of the medical term "schizophrenia" in this case, as it contributes to the widespread and harmful negative associations attached to mental illness (Castricano and Corman 2016, chap. 12), he was definitely on to something. He was on to something at least in the sense that to call such a disparity by the term that would probably sound more natural in this case, i.e. "hypocrisy," would be to wrongly assume that those guilty of the disparity in question are conscious of it, whereas most likely, they are not. They just do not see that their official stance and their actual practices are in tension.

Similarly, the subjects of psychological experiments who act according to the impression management mechanism (giving answers that present them in what they think is a better light) may not even be aware of that and give their answers entirely honestly as the mechanism may simply operate below the level of explicit consciousness (Schlenker 1980, 6). So, even if the subject is trying to be sincere, he or she may still misrepresent their own attitudes. In order to allow the subjects to give answers that would be both honest and true, researchers unfortunately have to be dishonest themselves. The standard practice is to deceive the participants about the true aim of the study, by presenting it as having a different aim, and at the same time avoiding the true aim being inferable from the very design of the study.

One way to achieve that goal, and the one we had chosen, is to ostensibly present a scale as measuring a construct X that is unrelated to the one it actually studies (let us call it 'Y'), and to mix the items from that scale with other questionnaire items such that the latter items do not stand out and the resulting questionnaire overall is believable as measuring the ostensible construct X. To do so, the items which serve as camouflage obviously have to be larger in number than the ones measuring the construct Y.

This is a good method in general. But applying it in practice in this case meant that a questionnaire that could successfully bury the 20 items which the AAS consists of would be definitely too long for most people to bear. So, given this and the aforementioned cultural specificity of the AAS, and that no other existing scale would work better, we developed our own scale, consisting of a smaller number of items. For various practical reasons, we settled on seven items, four of which we borrowed from the original scale developed by Herzog.[3] We had called the resulting scale Attitudes Toward Animal Welfare Scale (ATAW, for short), and this is how it looks:

1   The slaughter of whales and dolphins should be immediately stopped even if it means that some people will be put out of work.
2   The suffering of animals is an acceptable price for inventing drugs for humans.*
3   Human needs should always come before the needs of animals.*
4   I feel personally responsible for helping animals in need.
5   The low costs of food production do not justify maintaining animals under poor conditions.
6   Apes should be granted rights similar to human rights.
7   Basically, humans have the right to use animals as we see fit.*

In accordance with standard practice, we had included in the scale items that are reverse scored (those marked with an asterisk). We had also decided to let the subjects express their agreement with the statements on a seven-point scale, with answers ranging from "Completely disagree" (one point) to "Completely agree" (seven points).[4] Therefore, the total score that can be achieved by filling out the whole scale ranges between 7 and 49 points.[5] Analogically to the AAS, a higher score achieved on our scale is an indicator of pro-animal attitude.

Of course, in order to be certain that our scale accurately measures what it is supposed to measure, i.e. that it is valid, and that it is internally consistent, we had to conduct a pilot study first. Involving 55 participants ($N = 55$), it showed that the psychometric properties of the scale (its internal consistency and validity) were appropriate.[6] The validity was assessed by measuring how our scale correlates with the AAS. The result we achieved showed that the two correlate strongly and that the ATAW can be therefore said to measure what the AAS has been show to measure, that is, attitudes toward animal welfare, or how pro-animal one is.

Armed with such a scale we were then able to think about the experiments in which it could be used, and we came up with the idea of the following general design. The subjects from both the control and experimental group would first be informed that they are going to take part in a psychological study on a topic which has apparently nothing to do with animal welfare. In most of our experiments, the topic would be the relationship between the personality and the worldview of readers on the one hand and their perception of texts on the other. Then, the subjects in the experimental group are asked to read a story we hypothesize might have an impact on their relevant attitudes (call it "experimental story") and those in the control group are asked to read what we earlier called a narrative placebo, a narrative whose topic we considered neutral from the point of view of our study.

Having read their assigned stories, each group is then asked to complete a questionnaire (the same questionnaire in both cases), which consists of a few dozen questions that look like something that might measure the ostensible subject of the experiment (i.e. the relationship between the personality and the worldview of the reader on the one hand and his perception of texts on the other). Its first part is presented as measuring the personality and worldview in question, and includes items such as "I see myself as sympathetic, warm"[7] or "Cultural minorities should be supported and protected."

Interspersed with those items are the ATAW items, which we thought would look unexceptional in this company. After all, the ATAW items do look like something you might encounter in a questionnaire which measures personality (people do judge the personality of others on the basis of how they treat animals) and worldview (animal welfare is an ideological issue to most people). The reactions of the subjects that we later observed while performing our studies seem to have proven us right: no subject had expressed suspicion as to the real purpose of the study, something which does indeed happen in the case of experiments where less camouflage is involved.

The second part of the questionnaire would consist of items about the text itself. Some would be presented as measuring the extent to which the subjects remembered the content of the text (the name of the protagonist, etc.) and some as measuring the impressions it has generated. The latter items were borrowed from the so-called Transportation Scale, developed by M. C. Green and T. C. Brock (2000) in order to measure the extent to which people are immersed in the narratives they read.

While one purpose of this part of the questionnaire was to make our deception even more convincing, it was also to provide us with important data. Everybody will agree that a text read in a shallow way, either because of the lack of effort on the part of the reader or because it is for some reason too hard or boring for her, will work differently than a text which absorbs the reader thoroughly. It was therefore useful for us to control for the extent to which the subjects were transported into the story.

At the last stage, the subjects would fill out a set of metrical or demographic questions that provide information about their basic characteristics such as age, education, and the place of residence, all of which are variables that any psychologists should control in order to be able to assess how representative his or her results are. Finally, after all the subjects have filled out their questionnaires, they would be thanked for their cooperation and free to return to their non-experimental activities. The study would be over.

Now, everyone familiar with psychological experiments must have noticed that there is one important thing missing from the description of our procedure which is strictly related to the topic we have been concerned with in the preceding pages, that is deception. Since deception in psychological experiments is prima facie a breach of trust (an ethical shortcoming, that is), the researchers usually try to remedy that fault by informing the subjects immediately after the experiment is finished about its true purpose and by explaining why they had to be deceived. This procedure is called debriefing (Shaughnessy, Zechmeister, and Zechmeister 2012, 76–78).

While we agree that it is morally advisable to debrief one's subjects as soon as possible, we knew from the very beginning that we could not do this immediately in any of our experiments, and for some of them, we would have to wait for over a year. But if we decided to postpone debriefing for that long, this was justified by the same reason with which deception in psychological experiments is justified in the first place: by taking the precautions necessary for the results to be valid. We planned to conduct a whole series of experiments that involved thousands of people and social media. All of our subjects would be residents of the same country and at least half of them would live in the same, albeit large, city. In the age of Facebook and other social media, with the possibility for private individuals to spread information rapidly and widely, we felt it necessary to keep the purpose of our research secret until the last study was completed. This was accepted by the appropriate Ethical Committee but, of course, such committees are not morally infallible and sometimes make mistakes, so readers are welcome to make their own judgments about our decision too.

There Are Stories, and then There Are "Stories"

As soon as we had explained to ourselves what attitudes are and how best to study them, it might have seemed that we were ready to conduct our investigations. In particular, it might have appeared that we secured ourselves against the danger of chasing a chimera – searching for a hidden letter that does not exist. We knew clearly what we were looking for, didn't we? Not so fast. For we were about to study not only attitudes, but the impact that is exerted on them by *stories*, and "story" (or "narrative") is a notion that deserves some attention, too.

We can all agree that a story is an account of events. But accounts of events come in different shapes and flavors. For instance, *Babe* is a feature film, while *Black Beauty* is a novel, and there are many significant differences between how film and literary narratives work. And there are also stories told through other media, including music, the video game, the comic book, and the like. Certainly, it would be extremely difficult for us to arrive at experimental conclusions that would be valid for stories expressed in *any* medium. We had to narrow our focus, and given our expertise, we decided to focus on written narratives. But that qualification was still insufficient because written narratives include also texts such as the following:

> In order to provide a dosing comparison to a therapeutically relevant endpoint, both drugs were tested against amphetamine-induced disruption of prepulse inhibition as well. In the autoshaping task, rats were exposed to repeated pairings of stimuli that were differentially predictive of reward delivery. Conditioned approach to the reward predictive cue (sign-tracking) and to the reward (goal-tracking) increased during repeated pairings in the vehicle treated rats. Haloperidol and olanzapine completely abolished this behavior at relatively low doses (100 μg/kg). This same dose was the threshold dose for each drug to antagonize the sensorimotor gating deficits produced by amphetamine. At lower doses (3–30 μg/kg) both drugs produced a dose-dependent decrease in conditioned approach to the reward predictive cue. There was no difference between drugs at this dose range which indicates that olanzapine disrupts autoshaping at a significantly lower proposed DRD2 receptor occupancy. Interestingly, neither drug disrupted conditioned approach to the reward at the same dose range that disrupted conditioned approach to the reward predictive cue.
>
> (Danna and Elmer 2010)

It is a written account of a set of events, isn't it? But, leaving aside the question of the thematic content and focusing for the moment on the manner this story is told, we were definitely *not* interested in such dry, bare, and technical accounts. And there were plenty of other kinds of stories that we were not concerned with either. So *what* kind of narratives were we interested in? Consider *Black Beauty* and the other story that we mentioned as having a concrete social impact on people's concern for animal welfare –"They Die Piece by Piece" published in *The Washington Post*. What do these two have in common that enabled them to move their readers so profoundly?

One thing that definitely stands out is their literary quality. It is reasonable to suppose that they would not have succeeded, had their content been conveyed in a dry, scientific way rather than as it actually was, in a *literary* manner. But this qualification in itself still does not take us very far as literariness is a contested notion (Attridge 2017). While literature has been studied for thousands of years – far longer than attitudes, for instance – there is simply much less scholarly consensus on what it is than there is on attitudes. There is so little consensus that some authors declare that trying to define literature in general is a futile exercise; that it is a phenomenon "too amorphous to talk about" (Rorty 2005, 146). Indeed, so many utterly different things have been called "literature" across the ages and are called so today, that it is quite hard to tell what they all have in common. What is literature if you can call by that name both Tolstoy's mammoth novel *War and Peace* and poems which consist of a single blank page, including the untitled poem from the collection *The Death to Art* by Tolstoy's renowned compatriot, the futurist Vasilisk Gnedov?[8]

Faced with such a diversity of forms, the scholars who do not want to abandon the task of defining the notion of literature sometimes resort to the strategy of characterizing literature in terms of the response it generates. This is the strategy we adopt in this book, following such philosophers as John Dewey, Richard Shusterman, and Alan H. Goldman (Shusterman 1997, 2000a, 2000b; Dewey 2008; Małecki 2010; Koczanowicz and Małecki 2012; Goldman 2013). Goldman in particular sees literature as the kind of discourse that is aimed at engaging all our mental powers (including cognition, affect, perception, and imagination) to a heightened degree and in a concerted way. When a verbal work succeeds at this, "we can be said to lose our ordinary, practically oriented selves in the alternative world of the work. Our physical bodies disappear from consciousness, and our fully engaged mental capacities fully focus on this other world" (Goldman 2013, 82–83). Such a text is then a successful example of literature.

On this definition, a story with a literary quality is not necessarily one that belongs to a genre traditionally associated with the term "literature," like the novel, the short story, novella, or the epic poem. It may be a journalistic piece on, say, the discovery of the Higgs boson or on brain transplants in monkeys, provided that it is written in a way that engages our emotions, senses, imagination, and cognition. More than that, it does not even have to be published, or created by a professional or aspiring author. In fact, whenever a story engages our capacities in the way just described, it counts as literary, and whoever is apt at telling such stories can be described as possessed of a literary skill. We all know people who are not professional writers or even avid lovers of books, but whose accounts of events always seem to grab our attention. Whatever they talk about (a trip to a store, their kids' problems at school, or successes at work), we take pleasure in listening to them, are curious to know how the story ends, and can vividly imagine the events they talk about. They are, whether they know it or not, authors of literary stories.

One might even say that the most crucial capacity of a literary story-teller is to be able to turn any material into a tale that will induce the kind of aesthetic response we have been talking about. It is therefore no wonder that Marcel Proust – perhaps the greatest novelist of all time, and at the same time somebody who wrote extensively about such utterly prosaic details as lying in bed sleepless or eating a cake – seems to have honed his writing skills by retelling the most mundane stories he could find in "the news-in-brief section" of his favorite newspaper *Le Figaro* (De Botton 1997, 37). As Lucien Daudet reports: "A news-in-brief told by him turned into a whole tragic or comic novel, thanks to his imagination and fantasy" (quoted in De Botton 1997, 37).

By way of example, here are some excerpts from an issue of *Le Figaro* that Proust most likely had read, published in May 1914:

> At a busy crossing in Villeurbanne, a horse leapt into the rear carriage of a tram, overturning all the passengers, of whom three were seriously injured and had to be taken to hospital.
>
> While introducing a friend to the workings of an electric power station in Aube, M. Marcel Peigny put a finger on a high voltage cable and was at once fatally electrocuted.
>
> A teacher, M. Jules Renard, committed suicide yesterday in the Matropolitain, in the Republique station, by firing a single revolver shot into his chest. M. Renard had been suffering from an incurable disease.
>
> (quoted in De Botton 1997, 37)

And here is Alain de Botton's reconstruction of what Proust's retelling of the above stories might look like:

> Jules Renard? An unhappily married, asthmatic chemistry teacher employed by a Left Bank girls' school, diagnosed with colon cancer. The electrocuted Marcel Peigny? Killed while impressing a friend with a knowledge of electrical hardware in order to encourage a union between his harelipped son, Serge, and his friend's uncorseted daughter, Mathilde. And the horse in Villeurbanne? A somersault into the tram provoked by misjudged nostalgia for a show jumping career, or vengeance for the omnibus which had recently killed its brother in the market square, later put down for horse steak, suitable for feuilleton format.
>
> (1997, 37–38)

Of course, these are mere skeletons, or germs, of Proust's hypothetical novels, but when juxtaposed with the actual quotes from *Le Figaro*, they already demonstrate the difference between stories that merely recount events, and literary stories that we insist on. Even though they are as short as the newspaper extracts, and can be described as only potentially literary, there is something about them that incites our imagination, emotions, associations, and curiosity to a higher degree than in the case of the paper's accounts. We feel transported to the world in which those poor beings lived and died, and we want more of this experience, whereas we could easily pass over the original newspaper notices. When we deal with full-blown literary stories, not their germs such as the above, such impressions are still heightened.

Recall the readers of Rousseau's *Julie* who cried actual tears over the plight of the novel's heroine, or were so absorbed in reading that they seemed to lose the sense of where their identity ended and hers began (Hunt 2007; Oatley 2011, 168–69). According to historical accounts, a similar experience had befallen the readers of *Black Beauty* whose anger and protest led later to actual animal welfare reforms. But as we already know, novels are not the *only* animal stories that are written in such a manner that they can move us deeply. Let us come back to the journalistic story "They Die Piece by Piece" and consider the scene with which it opens:

> It takes 25 minutes to turn a live steer into steak at the modern slaughterhouse where Ramon Moreno works. For 20 years, his post was "second-legger," a job that entails cutting hocks off carcasses as they whirl past at a rate of 309 an hour.
>
> The cattle were supposed to be dead before they got to Moreno. But too often they weren't.
>
> "They blink. They make noises," he said softly. "The head moves, the eyes are wide and looking around."
>
> Still Moreno would cut. On bad days, he says, dozens of animals reached his station clearly alive and conscious. Some would survive as far as the tail cutter, the belly ripper, the hide puller. "They die," said Moreno, "piece by piece."
>
> (Warrick 2001)

If reading these passages feels like jumping into a pit of horror, this is definitely not only due to the content itself but also thanks to the way it was arranged and expressed. Take the journalist's use of *paronomasia*, a figure of speech where two similarly sounding, but etymologically unrelated words, are juxtaposed in order to suggest an intriguing connection ("steer" and "steak").[9] Or consider the use of contrast (the noises made by the animals and the soft voice of the slaughterer) and the onomatopoeia of the staccato rhythm of the last two sentences (strengthened by the accumulation of explosive consonants *p*, *t*, *b*, *d*), which mimics the automatic process of killing a live creature piece by piece. And what about all the swirling carcasses, moving heads, rolling eyes, and ripped bellies? The picture is almost nauseating; and again not only because of the suffering it depicts, but also because all of its elements seem to be constantly spinning in different directions. And all this has been conjured by just a few carefully written sentences. It is precisely texts like this that made us focus on animal stories with a *literary* quality.

Another reason, aside from the fact that literary stories can make any content absorbing, is that they can make any topic attractive (Strange 2002). This is important when you consider how many people simply do not *want* to learn about the details of animal oppression. They do not want to see them or hear about them, either because they are uninterested or because they sense that this would cause them too much distress and possibly guilt (Taylor 2016). If they *were* generally interested, then pro-animal organizations would not have to resort to scandalizing, or even to paying, the members of the public to turn their attention to material that might hopefully change their attitudes. PETA's campaign "Holocaust on Your Plate," involving billboards juxtaposing the pictures of concentration camp prisoners with the pictures of farm animals is one example (Deckha 2008; Munro 2012; Buettner 2016, chap. 6). Another would be the recent project of the American organization Farm Animal Rights Movement, which tours the States with their mobile video truck, offering a dollar to anyone willing to watch a graphic 4-min documentary titled "10 Billion Lives" (Elist 2012).

Literary stories help to circumvent such problems since literature can make any topic attractive, no matter how repulsive, boring, or blasphemous it might otherwise seem to its audience. For instance, it is safe to presume that among the readers of Donna Leon's *New York Times* bestseller *Beastly Things*, a detective novel dealing with the horrors of meat production, there are people who would vehemently reject an invitation to watch documentary footage on that topic. Yet, they simply had to learn a lot about it if they wanted to follow the main protagonist of the novel, Commissario Guido Brunetti. They were left no alternative by the construction of the plot, whose trajectory led them through places such as this one:

> Six or seven yellow-booted men in white rubber coats and yellow hard-hats moved below [Brunetti and Vianello] in the cement-floored cubicles and did things with knives and pointed instruments to pigs and sheep. Animals fell at the feet of the men, but some managed to flee, crashing into the walls before slipping and falling. Others, wounded and bleeding and unable to get to their feet, continued to flail about with their legs, feet scrambling against floors and walls, while the men dodged their hooves to deliver another blow.
>
> Some of the sheep, Brunetti noticed, were protected from the knives by their thick coats and had to be struck repeatedly on the

head by what looked like metal rods that ended in hooks. Occasionally, the hooks were used for other purposes, but Brunetti looked away before he could be sure of that, though the wail that always followed the desertion of his eyes left no doubt about what went on.

The sheep made low, animal noises – grunts and bleats – while the pigs struck him as sounding not unlike what he or Vianello would sound like, were they down there and not up here. The calves bleated.

(Leon 2012, 152–53)

Finally, literary stories are also well-suited to providing vivid, emotionally charged accounts of individual suffering, which is important here because such accounts are highly effective in raising public awareness of large-scale misery, particularly in comparison with statistical or numerical data, on which animal welfare campaigns often rely (Slovic 2007; Bal and Veltk amp 2013). In fact, when the public is exposed to such data, this often results in so-called psychophysical numbing, i.e. a collapse of compassion (Slovic 2007; Slovic and Slovic 2015, pt. 1).

This brings us to one final criterion that we imposed on the texts we wanted to study. We wanted them to be stories that portrayed the *plight* of non-human animals, and where these conditions were inflicted by people *unnecessarily*. That criterion of course meant that we consciously left a plethora of animal stories beyond the scope of our study, including some that might potentially have a positive impact on people's concern about the well-being of animals. It is possible, after all, that these same effects might result from reading narratives about remarkable cases of positive human–animal bonds or about the way animals behave in their natural habitat.

However, we wanted to keep to the kind of material that has been historically reported to influence attitudes toward animal welfare, and this material happened to always include depictions of unnecessary animal suffering (Pearson 2011; cf. Eisenman 2013).[10] Moreover, we wanted to study it as thoroughly as our resources allowed. Other researchers are welcome to explore the other kinds of literary animal narratives through their own experiments, and we ourselves may do that, too, in the future.

Why Is It So Hard to Experiment with Literature?

If it is their capacity to induce aesthetic experiences that makes literary stories so impactful, then it is also something that makes studying that impact through experiments extremely difficult (Kaufman and Libby 2012; Johnson et al. 2013; Johnson, Huffman, and Jasper 2014; Vezzali et al. 2015). Here, we should remind ourselves that, however useful, the experimental method nevertheless has vices, and some of these are related to its main virtue, the level of control it affords the researcher.

These vices are more pronounced in the case of some objects of study than others, and they are unfortunately particularly pronounced in the case of the psychological effects of literary reading. It is for this reason that much existing research into such effects of literature suffers from various methodological problems. No matter how fine a scholar you are, you are bound to make some tough methodological choices that will often negatively affect your results.

Consider, first, that in order to study the effects of literary reading experimentally, the best choice is to conduct your study in a laboratory space. As our example of the vitamin C study was meant to show, it is the laboratory, after all, which allows for the greatest level of control possible (Shaughnessy, Zechmeister, and Zechmeister 2012, chap. 6; cf. Webster and Sell 2007). It is only natural, then, that this is where most scholars perform their experiments on readers of literary texts. But the problem is that, as we have already agreed, reading literature is very different from taking a vitamin pill.

For instance, the mere space *where* we take a pill is largely irrelevant to its effects, and thus while drug leaflets may advise us *when* we should take a pill and *what* we should take it with, they remain silent on the place where the pills are to be taken. It does not matter whether we do so in the silence of our homes or in the bustle of a busy street. This is not the case with reading literature – at least for most people. The kind of aesthetic experience that is definitional of literary reading demands sustaining a heightened level of mental focus for a period of time, and to achieve this most readers need a space that will not only be quiet, and where they will be uninterrupted, but will also afford them a feeling of comfort and intimacy (Burke 2011, 101). Granted, being in an alien, laboratory space and aware of the fact that we are the subject of a study can ruin such a feeling for some, something which may negatively affect the results. That is, such results may not have much to do with how people read in the real world and a fortiori with how they are influenced by what they read.

Still another problematic consequence of studying literary reading in a lab is that this reduces the possible time that can be spent on this activity and therefore the kinds of texts which can be studied. It would be practically impossible to make somebody read *War and Peace*, or any other novel for that matter, in an environment sufficiently controlled by the experimenter so as to preclude any external factors significantly interfering with the results. The texts studied by experimenters therefore tend to be relatively short and for this reason do not constitute a representative sample of the stories people read in real life.

Their length, however, is not the only problem that texts pose for those who want to study their impact experimentally. There are also problems related to their other formal features, as well as their content. Say, you want to study the influence of animal stories on attitudes toward animals and their welfare, like we do, and you find what you think is a suitable story. But while the text involves the motif of harm being done to animals by humans, it also depicts the plight of a human being. This is a disadvantage insofar as you then cannot be sure whether, if the impact is observed, it is due to the animal component of the story or perhaps due to the plight of the human being, which might have aroused sympathetic feelings in the subjects to such an extent that this also affected their concern for animals.

Or imagine that while your text does not additionally depict the plight of human beings, it depicts the human being who inflicts suffering upon an animal in a glamorizing way, as it often happens in Hemingway's prose (2012), for instance. If your results were insignificant, would you then be entitled to say that stories do not influence attitudes toward animal attitudes? Wouldn't you be left wondering what the result would have been, had the human perpetrator been portrayed less sympathetically?

These are but two possible examples of a feature of the text which might hamper your results, and there are many such features possible for most experiments. Faced with such a challenge you basically have two options to choose from. Either you try to find another suitable text or excise from the one you have those components that are undesirable. Neither option is guaranteed to succeed. Sometimes a suitable text is hard to find, and sometimes the undesirable features of the narrative are so intertwined with its other features, that they cannot be extracted. It is therefore hardly surprising that sometimes experimenters choose to write experimental texts themselves, taking good care *not* to include in those texts anything that might render the results inconclusive (cf. Kaufman and Libby 2012, 9). The result is a purified textual product designed to fit the purpose of their study.

But note that whereas it is fairly easy to isolate one active ingredient in a pill and consider the other ingredients as entirely irrelevant to the effect we want to study, this is not the case with literary narratives. The texts prepared by experimenters may possess the purity of form and content necessary for the experimental results to be unambiguous, but it is hard to resist the impression that such textual products are hardly like the actual literary stories people read. Leaving aside the trivial fact that most researchers do not possess genuine literary skills, the problem is that what makes a story a *literary* story, and a fortiori a good literary story (one that people want to read), is precisely its richness, ambiguity, all the unnecessary details that get in the way of an experimenter (Barthes 2010). Get rid of these supposedly unnecessary details and you will get rid of its literary essence. This is why the texts prepared by researchers themselves are often referred to as "textoids," rather than texts (Schmalhofer and Perfetti 2007, 19).

Finally, there is the all too common logistical problem of finding a sufficiently large and diverse sample of subjects for one's experiments.

Whereas research on pills has good chances of receiving funding from governments and pharmaceutical companies that is sufficient for securing such a sample, research on literature usually cannot count on support on this grand scale. And so, psychologists who study literature are forced to resort to the pool of subjects that is the most easily and least costly available to them, that is, university students (Hakemulder 2000; Bal and Veltkamp 2013). This drastically limits the potential of their results to be extrapolated to the so-called man, or woman, in the street. Students as a group are just too specific for that – in terms of their age, economic background, not to mention education.

But to be fair, it needs to be mentioned that that kind of limitation applies to most experimental research in psychology as most of the topics that psychologists study, not only literature, are thought to deserve much less funding than pills. Hence, the title of a recent iconoclastic article on psychological research, "The Weirdest People in the World?," which aims to challenge the widespread presumption that psychologists study the psychology of people in general (Henrich, Heine, and Norenzayan 2010). No, as a matter of fact, psychologists study the psychology of people who are quite WEIRD in comparison to the population of the world, with WEIRD standing for Western, Educated, Industrialized, Rich, and Democratic societies. But recall that the people whom psychologists study are also quite weird in comparison to the population of their own respective countries. As one author caustically observed, if the journals in which psychologists publish "were renamed to more accurately reflect the nature of their samples," their titles would have to be "*The Journal of Personality and Social Psychology of American Undergraduate Psychology Students*," and such like (Henrich, Heine, and Norenzayan 2010, 63).

To sum up, attempts to exert strict control in experiments such as the above often result in a decrease in what psychologists call the ecological validity of the study, i.e. the extent to which its results can be generalized (or extended) "to real-world, everyday settings" (Kellogg 2002, 20). It is therefore easy to understand that in taking up the task of providing reliable empirical data on the impact of stories on attitudes toward animals, we faced a considerable challenge. On the one hand, we wanted to study the subject through controlled experiments. On the other, we did not want to study weird people, or at least not exclusively. In particular, we did not want to study weird people reading weird texts (i.e. textoids that have little to do with narratives we read every day) in weird, artificial circumstances. We tried to face the challenge by doing the following.

First, we avoided using textoids, choosing instead a variety of texts that not only have been penned by actual writers but are claimed to have had the kind of impact we want to study. They are, among others: Marshall Saunders's *Beautiful Joe*, the Canadian equivalent of *Black Beauty* and the first novel by a Canadian writer "to sell a million copies" (Davis 2016, 67; cf. Fiamengo 2012; Saunders 2015); Alice Walker's "Am I Blue?", again a staple of literary animal studies and a text widely read also outside of that context (2010; cf. Hooker 2005); Dostoyevsky's famous depiction of horse beating from *Crime and Punishment*; a fragment of Gail Eisnitz's *Slaughterhouse*, a journalistic book which was recommended in the *Vegan & Vegetarian FAQ* as a "great resource" of narratives that can "hook the readers' attention" (Breier and Mangels 2001, 46); "The Dead Body and the Living Brain," a fine example of literary journalism penned by one of the genre's most famous figures Oriana Fallaci (2010) and anthologized in the collection *Other Nations*.

Of course, it is not that all of these stories fit our designs perfectly. But all of them were *generally* suitable, and we decided that it is better to manipulate them slightly in cases where that was necessary than to write our own narratives from scratch. The reader will be dutifully informed about each and every one of our interventions in the original texts, to see whether they were rather gentle face-lifts or acts of literary butchery.

Second, thanks to a generous grant from the National Science Center, we were able to pursue our project on a relatively grand scale and test our hypothesis on thousands of people of different backgrounds and age. Whenever our subjects recruited from university students, we made sure that they come from different departments and different types of institutions of higher education. Some of them studied journalism, some computer graphics, and some physiotherapy. Some came from universities and some from technical schools. We had also conducted studies on high school students from schools of different profiles, both science- and humanities-oriented. And finally, hundreds of our subjects were not students at all. Those people had basically one thing in common, they liked to read books, and apart from that they differed among each other as much book readers in general do. In their wealth, in where they lived, their profession, and the like. In other words, they were not weird people at all.[11]

What is more, the latter group of people participated in experiments that took place outside of laboratory spaces. They did not read texts in circumstances that would be weird for them, but in a place of their own choosing, one that, we assume, they felt to be natural. Those studies indeed belonged to the category of experiments that psychologists sometimes call "natural," which do not allow for strict control of variables in the study, but are possessed of a much higher level of ecological validity than laboratory studies (Dunning 2012). In order to balance for the lack of strict control in those studies, we also conducted a number of laboratory experiments. This way, we could take advantage of the virtues of both types of experiments, and at the same neutralize their vices by comparing their results. Sometimes you can both have your cake and eat it.

Truths, Damned Truths, and Statistics

It should be clear by now that the data we obtained in our experiments came at a significant price. First, there was the ethical cost of deception. Second, there was the effort invested by both the experimenters and the subjects. Third, there were also the actual financial costs on the part of the governmental agency that provided us with funding. However, despite all these costs, the numerical data contained in the questionnaires filled out by our subjects was of little cognitive value *in itself*. That data were *raw* material and had to be refined in order to transform it into psychological knowledge, that is justified true statements about how human minds work. Such a process of refinement is typically performed with the use of many complex statistical tools, and we had used some of those ourselves.

However, in what follows we will abstain from describing them all in detail. In particular, we will not include here the detailed statistics of all the effects, including interactive effects and post-hoc effects. If terms "interactive effects" and "post-hoc effects" have left some of the readers baffled, then this only proves that our decision was right. These are specialist terms and in order to explain their meaning to a reader unfamiliar with the methodology of the social sciences, we would have to write a whole book – a book entirely different from the one we want to write (one about stories, attitudes, and animals).[12] Moreover, providing these details in the case of each result we describe would be cumbersome and reduce the readability of the book even to the specialists.[13]

But there are still two basic statistical notions that have to be explained here, as they are necessary for the understanding of our main results. They are conventionally designated by letters p and η^2 (eta squared). Let us begin with the former as that inconspicuous letter p is in fact of great importance in psychological studies, standing for statistical significance. If the value of p is smaller than 0.05, this means that the result is statistically significant, and that we are entitled to say that our hypothesis has been confirmed (Shaughnessy, Zechmeister, and Zechmeister 2012, 385–86). The smaller the value of p, the smaller the probability that our results are due to a random error or chance. Strictly speaking, $p = 0.05$ means that we can assume that there is a 95% probability that our result is not due to chance or error. With p equaling 0.01, the probability would be 99%, and so on.

Why is 0.05 p the threshold of statistical relevance? This is a good question. Consider the fact that the difference between, say, $p = 0.04$ and $p = 0.06$ is obviously minimal. In the former case, we assume that there is a 96% probability in the other that the probability is 94%. However small the difference between them, one of these results will be judged by the scientific community as statistically relevant (and confirming the hypothesis), and the other will not.

One should remember here, though, that the judgment will not stem from any principle that is carved in stone. The assumption that $p = 0.05$ is a line dividing the realm of what is statistically significant from what is not should be seen rather as of conventional nature: an "arbitrary" choice necessitated by the fact that the line had to be drawn somewhere and that at the same time it could not be drawn where we would most like to draw it, at 100%, as the probability of random error can never be ruled out (Lindgren 1993, 303). In thinking about the p value and statistical relevance, one should also remember two further things. First, they are tightly linked to the number of participants in a study. The more of them, the smaller the risk of error, and the easier to obtain a statistically relevant result (Shaughnessy, Zechmeister, and Zechmeister 2012, 388–89). Second, a good (that is, small) value of p does not mean by itself that the effect it describes is large, or, in other words, that a given phenomenon is strong.

In experimental sciences, the strength of the phenomenon, or the size of the effect, is measured with the help of various values; but in this book, we decided to use only one of them, that which is conventionally designated with the second of the letters we mentioned above, that is η^2 (Shaughnessy, Zechmeister, and Zechmeister 2012, 402–3). As is generally agreed (though, like in the case of the p value, this is again a matter of convention), the small, the medium, and the large size of the effect is described by the values 0.1, 0.3, and 0.5 (cf. Cohen 1988, 7–14). To illustrate what "small" means here, let us come back to the example of impact of vitamin C on the symptoms of common cold. For instance, if in a study conducted on 500,000 subjects it was shown that taking vitamin C throughout the duration of the illness shortens it by 5 min, then while the result would be statistically relevant (the hypothesis was confirmed), it would have no practical meaning. The size of the effect would be small.

In reading the descriptions of our results below, the reader may then be surprised to learn that in the case of most of those that are statistically relevant, the size of the effect is indeed small. The reader's surprise will only grow, we suppose, if we tell him, or her, that this is the norm, not the exception, in social sciences, including social psychology. This means that for almost every psychological study that the reader knows from the press and which announces in sensationalist terms that "the researchers have shown" this or that, the phenomenon described therein is usually relatively weak. Very often it is the equivalent of the duration of common cold having been shortened by 5 min.

What does this mean for our main hypothesis (that stories can influence attitudes toward animal welfare) and does that not stand in tension with the example with which this book began: the enormous social influence of *Black Beauty*? The answer will be possible only after our results are fully presented, and we of course will dutifully give it then.

At this point, we can at least state that the experimental procedures we employed and the kind of statistical apparatus we applied to our raw data guarantee that our results are statistically significant, and therefore empirically sound. From now on, researchers talking about the psychological influence of animal narratives will have something more to rely on than philosophical speculation and anecdotes, personal or historical. Since this statement may sound a bit hard-nosed to our humanist readers, some of whom might have even had the unpleasant experience of having a social or natural scientist look down at them as insufficiently scientific and serious, we would like to end this chapter by making clear our position on the relation between the social sciences and the humanities.

First, while we do think that the humanities in general, not only those concerned with stories, would profit from adopting empirical methods to a greater degree, we do not think that they should imitate the social sciences in everything they do. We do not think that the social sciences have found a methodological holy grail when it comes to studying cultural and social phenomena such as stories. We also do not think that, as they currently function, the social sciences are the epitome of scholarly soundness which the humanities should look up to. The latter statement would be particularly difficult to maintain now, in the light of recent revelations that the results of a large percentage of social scientific studies either cannot be replicated, or can only be replicated in the way they were produced, that is, by tinkering with the obtained data (Pashler and Harris 2012; Collaboration 2015).

We would like to address two of the unfortunately quite widespread ways of tinkering with data, as our distancing ourselves from them has some consequences for the way our own results are presented. One of these ways lies in performing various statistical tricks on the data that allow the researcher to hold that the result falls below the magical 0.05 value of p while it in fact does not. This "occurs when researchers try out several statistical analyses and/or data eligibility specifications and then selectively report those that produce significant results" (Head et al. 2015). Some of these tricks are hard to detect as they involve quite sophisticated instruments such as "excluding, combining, or splitting treatment groups postanalysis, including or excluding covariates postanalysis," and the like (Bakker, van Dijk, and Wicherts 2012; Head et al. 2015). The scale of this practice can be indirectly illustrated by the fact that it even has its own popular names: "p-hacking" and "p-fishing." Poor Mark Twain did not even know how right he was when he famously quipped about lies, damned lies, and statistics! But of course, this is satire. Statistical methods still remain the best way to analyze empirical data, and the interference we describe is not inherent to them. Statistical analysis is merely a tool and as such can be used for various purposes, and we assure the reader that no statis-*tricks* were employed during the making of this book.

The other way to tinker with the data is much less sophisticated than *p*-hacking, but no less powerful, and involves drawers. We are quite serious. At least as serious as the prominent American psychologist Robert Rosenthal who invented the phrase "the file drawer problem" to describe the apparently widespread practice of revealing only the results of those experiments that confirm a given hypothesis (i.e. the results that are statistically significant) while keeping those that disconfirm it in … the file drawer indeed (Rosenthal 1979).

Imagine you have a very nice hypothesis and you perform 12 experiments to prove it right, but unfortunately four of those disconfirm it. What a shame! You have put so much effort and money in your project, after all. But you still have those eight experiments that worked, right? Do you have to tell anyone else about the other four? We are sure that our readers would act honestly in such circumstances, but unfortunately, some researchers yield to the temptation not to tell, aware that inconclusive or negative results are harder to publish than positive ones and afraid how the bodies that granted money for their research will react to such a failure.

In any event, we are happy to say that no file drawer was abused during the making of this book. We present in it even those experiments whose results were not as we hoped them to be. Some of our studies confirm our hypotheses, some do not, and we tell it like it is. One consequence of this is that the final picture painted by our results is not nice and simple. But on the other hand, we are sure that what our readers want is truth and nothing but the truth, even if it is difficult and complex. Besides, those kinds of damned complicated truths tend to be more interesting than damned easy lies.

Notes

1 Of course, there are numerous other ways to measure attitudes toward animals than the AAS. For a good overview, see, e.g. Serpell (2004).
2 Note that it was only two years after we had begun working on our project that Herzog published a shorter version of the AAS. See Herzog, Grayson, and McCord (2015).
3 In case the readers worry whether Harold Herzog might be upset about our butchering his scale and then stitching its parts with something else, they should be advised that Herzog, in the spirit of scientific collegiality, had offered his scale for other researchers to use for whatever purpose they like and explicitly stated that they are "welcome to modify it to meet their research needs." See "Animal Attitudes Scale," electronic document published at "Homepage for Harold Herzog" (2017), cf. Herzog, Grayson, and McCord (2015).
4 Here is the complete list of the possible answers: "Completely disagree" – 1; "Disagree" – 2; "Somewhat disagree" – 3; "Neither agree nor disagree" – 4; "Somewhat agree" – 5; "Agree" – 6; "Completely agree" – 7.
5 7 questions marked on a 7-point scale. Note that hereafter we will report average scores, i.e. a total score divided by seven questions.

6 To put it in more technical terms, the internal consistency of the scale measured with so-called Cronbach's α was 0.81, while its validity was $r = 0.7$, $p < 0.0001$.

7 This and nine other items in our questionnaire which directly concerned personality were taken from the so-called Ten-Item Personality Inventory. See Sorokowska et al. (2014).

8 The problem of amorphousness, by the way, pertains not only to literature, but all art in general. What, for instance, is the common denominator of Leonardo da Vinci's "Last Supper" (an extraordinary painting created with the artist's own hands) and Marcel Duchamp's readymades: i.e. ordinary objects such as urinals that the author did not create himself but bought in a store and then presented in a gallery? And why do we apply the term 'musical work of art' to both Mozart's *The Marriage of Figaro* (which, at least according to Emperor Joseph II, contains "too many notes") and John Cage's "4,33", which consists of no notes whatsoever, but rather 4 min and 33 s of silence? For an overview of theoretical debates on the nature of art in contemporary aesthetics, see, e.g., Shusterman (2000b) and Małecki (2010).

9 Wales (2011, 287).

10 That the suffering depicted in these stories was unnecessary is quite important here given that there are good reasons to think that it is mainly undeserved suffering that provokes sympathetic reactions. For more on this point, see Sklar (2013).

11 They were not weird at least in comparison to the population of their own country, even if they were WEIRD in the sense of belonging to the western world.

12 Cf. Shaughnessy, Zechmeister, and Zechmeister (2012, pt. 5).

13 So as far as statistical details are concerned, we will only state here generally, for all those interested, that the results presented in our book had been described on the basis of analyses of variance (ANOVA) with post hoc analyses (LSD test) for all experiments. In the case of longitudinal experiments, in Chapter 7, we analyzed the results using ANOVA with repeated measures. The non-specialists readers should also know that these days, most such analyses are performed with the help of software and that this is how we worked too: we obtained our results by means of Statistica Software, version 1.2.

Works Cited

Attridge, Derek. 2017. *The Singularity of Literature*. Routledge Classics. Abingdon; New York: Routledge.

Bakker, Marjan, Annette van Dijk, and Jelte M. Wicherts. 2012. "The Rules of the Game Called Psychological Science." *Perspectives on Psychological Science* 7 (6): 543–54. doi:10.1177/1745691612459060.

Bal, P. Matthijs, and Martijn Veltkamp. 2013. "How Does Fiction Reading Influence Empathy? An Experimental Investigation on the Role of Emotional Transportation." *PLoS ONE* 8 (1): e55341. doi:10.1371/journal.pone.0055341.

Barthes, Roland. 2010. "The Reality Effect." In *The Rustle of Language*. Translated by Richard Howard, 141–48. Berkeley: University of California Press.

Becker, Kate. 2014. "The Mistaken Assumptions That Changed Physics History." Accessed June 18, 2014. www.pbs.org/wgbh/nova/blogs/physics/2014/06/the-assumptions-physicists-make/.

Breier, Davida Gypsy, and Reed Mangels. 2001. *Vegan & Vegetarian FAQ: Answers to Your Frequently Asked Questions*. Baltimore, MA: Vegetarian Resource Group.

Buettner, Angi. 2016. *Holocaust Images and Picturing Catastrophe: The Cultural Politics of Seeing*. London: Routledge.

Burke, Michael. 2011. *Literary Reading, Cognition and Emotion: An Exploration of the Oceanic Mind*. Routledge Studies in Rhetoric and Stylistics 1. New York: Routledge.

Castricano, Jodey, and Lauren Corman. 2016. *Animal Subjects 2.0*. Waterloo, ON: Wilfrid Laurier University Press.

Flynn, Clifton P. 2003. "A Course Is a Course, of Course, of Course (Unless It's an Animals and Society Course): Challenging Boundaries in Academia." *International Journal of Sociology and Social Policy* 23 (3): 94–108. doi:10.1108/01443330310790273.

Cohen, Jacob. 1988. *Statistical Power Analysis for the Behavioral Sciences*. Hillsdale, NJ: L. Erlbaum Associates.

Collaboration, Open Science. 2015. "Estimating the Reproducibility of Psychological Science." *Science* 349 (6251): aac4716. doi:10.1126/science.aac4716.

Crano, William D., and Radmila Prislin, eds. 2008. *Attitudes and Attitude Change*. Frontiers of Social Psychology. New York; London: Psychology Press.

Danna, C. L., and G. I. Elmer. 2010. "Disruption of Conditioned Reward Association by Typical and Atypical Antipsychotics." *Pharmacology, Biochemistry, and Behavior* 96 (1): 40–47. doi:10.1016/j.pbb.2010.04.004.

Davis, Janet M. 2016. *The Gospel of Kindness: Animal Welfare and the Making of Modern America*. Oxford ; New York: Oxford University Press.

De Botton, Alain. 1997. *How Proust Can Change Your Life: Not a Novel*. New York: Pantheon Books.

Deckha, Maneesha. 2008. "Disturbing Images: Peta and the Feminist Ethics of Animal Advocacy." *Ethics & the Environment* 13 (2): 35–76.

Dewey, John. 2008. *The Later Works of John Dewey, 1925–1953: 1934, Art as Experience*. Edited by Jo Ann Boydston and Abraham Kaplan. Carbondale: Southern Illinois University Press.

Dunning, Thad. 2012. *Natural Experiments in the Social Sciences: A Design-Based Approach*. Cambridge; New York: Cambridge University Press.

Eagly, Alice Hendrickson, and Shelly Chaiken. 1993. *The Psychology of Attitudes*. Fort Worth, TX: Harcourt Brace Jovanovich College Publishers.

———. 1998. "Attitude Structure and Function." In *The Handbook of Social Psychology*, edited by Daniel Gilbert, Susan Fiske, and Lindzey Gardner, 269–322. New York: Oxford University Press.

Eisenman, Stephen. 2013. *The Cry of Nature: Art and the Making of Animal Rights*. London: Reaktion Books Ltd.

Elist, Jasmine. 2012. "FARM Campaign Pays Viewers $1 to Watch Graphic Anti-Meat Video." *Los Angeles Times*, May 25, 2012. http://articles.latimes.com/2012/may/25/local/la-me-gs-campaign-pays-viewers-1-to-watch-graphic-antimeat-video-20120525.

Fallaci, Oriana. 2010. "The Dead Body and the Living Brain." In *Other Nations: Animals in Modern Literature*, edited by Tom Regan and Andrew Linzey, 117–24. Waco, TX: Baylor University Press.

Felski, Rita. 2015. *The Limits of Critique*. Chicago: The University of Chicago Press.

Fiamengo, Janice Anne. 2012. *Other Selves: Animals in the Canadian Literary Imagination*. Ottawa: University of Ottawa Press. http://muse.jhu.edu/books/9780776617701/.

Francione, Gary L. 2000. *Introduction to Animal Rights: Your Child or the Dog?* Philadelphia, PA: Temple University Press.

Goldman, Alan H. 2013. *Philosophy and the Novel*. Oxford: Oxford University Press.

Green, Melanie C., and Timothy C. Brock. 2000. "The Role of Transportation in the Persuasiveness of Public Narratives." *Journal of Personality and Social Psychology* 79 (5): 701–21.

Hakemulder, Jèmeljan. 2000. *The Moral Laboratory: Experiments Examining the Effects of Reading Literature on Social Perception and Moral Self-Concept*. Utrecht Publications in General and Comparative Literature, v. 34. Amsterdam; Philadelphia, PA: J. Benjamins Pub.

Head, Megan L., Luke Holman, Rob Lanfear, Andrew T. Kahn, and Michael D. Jennions. 2015. "The Extent and Consequences of P-Hacking in Science." *PLOS Biology* 13 (3): e1002106. doi:10.1371/journal.pbio.1002106.

Hemingway, Ernest, and Séan A. Hemingway. 2012. *Hemingway on Hunting*. First Scribner Classics hardcover edition. New York: Scribner Classics.

Henrich, Joseph, Steven J. Heine, and Ara Norenzayan. 2010. "The Weirdest People in the World?" *The Behavioral and Brain Sciences* 33 (2–3): 61–83; discussion 83–135. doi:10.1017/S0140525X0999152X.

Herzog, Harold, Nancy S. Betchart, and Robert B. Pittman. 1991. "Gender, Sex Role Orientation, and Attitudes Toward Animals." *Anthrozoos: A Multidisciplinary Journal of the Interactions of People & Animals* 4 (3): 184–91. doi:10.2752/089279391787057170.

Herzog, Harold A., and Steve Mathews. 1997. "Personality and Attitudes Toward the Treatment of Animals." *Society and Animals* 5 (2): 169–75.

Herzog, Harold, Stephanie Grayson, and David McCord. 2015. "Brief Measures of the Animal Attitude Scale." *Anthrozoös* 28 (1): 145–52. doi:10.2752/089279315X14129350721894.

"Homepage for Harold Herzog." 2017. January 24, 2017. http://paws.wcu.edu/herzog/.

Hooker, Deborah Anne. 2005. "Reanimating the Trope of the Talking Book in Alice Walker's 'Strong Horse Tea.'" *The Southern Literary Journal* 37 (2): 81–102. doi:10.1353/slj.2005.0018.

Hunt, Lynn. 2007. *Inventing Human Rights: A History*. 1st ed. New York: W.W. Norton & Co.

Johnson, Dan R., Brandie L. Huffman, and Danny M. Jasper. 2014. "Changing Race Boundary Perception by Reading Narrative Fiction." *Basic and Applied Social Psychology* 36 (1): 83–90. doi:10.1080/01973533.2013.856791.

Johnson, Dan R., Daniel M. Jasper, Sallie Griffin, and Brandie L. Huffman. 2013. "Reading Narrative Fiction Reduces Arab-Muslim Prejudice and Offers a Safe Haven From Intergroup Anxiety." *Social Cognition* 31 (5): 578–98. doi:10.1521/soco.2013.31.5.578.

Kant, Immanuel. 1993. "On a Supposed Right to Lie Because of Philanthropic Concerns." In *Grounding for the Metaphysics of Morals: With On a Supposed*

Right to Lie Because of Philanthropic Concerns, translated by James W. Ellington, 63–68. Indianapolis, IN: Hackett Publishing.

Kaufman, Geoff F., and Lisa K. Libby. 2012. "Changing Beliefs and Behavior through Experience-Taking." *Journal of Personality and Social Psychology* 103 (1): 1–19. doi:10.1037/a0027525.

Kellogg, Ronald T. 2002. *Cognitive Psychology*. London: SAGE Publications.

Koczanowicz, Dorota, and Wojciech Małecki. 2012. *Shusterman's Pragmatism: Between Literature and Somaesthetics*. Rodopi.

Lee, Kibeom, Michael C. Ashton, Julie Choi, and Kayla Zachariassen. 2015. "Connectedness to Nature and to Humanity: Their Association and Personality Correlates." *Frontiers in Psychology* 6 (July). doi:10.3389/fpsyg.2015.01003.

Leon, Donna. 2012. *Beastly Things*. Commissario Guido Brunetti Mystery. New York : Atlantic Monthly Press.

Lindgren, B. W. 1993. *Statistical Theory*. 4th ed. New York: Chapman & Hall.

Loeb, Abraham. 2014. "Benefits of Diversity." *Nature Physics* 10 (9): 616–17. doi:10.1038/nphys3089.

Maio, Gregory R., and Geoffrey Haddock. 2012. *The Psychology of Attitudes and Attitude Change*. Los Angeles, CA: Sage.

Małecki, Wojciech. 2010. *Embodying Pragmatism Richard Shusterman's Philosophy and Literary Theory*. Frankfurt am Main; New York: Peter Lang.

Muller, John P., and William J. Richardson, eds. 1987. *The Purloined Poe: Lacan, Derrida, and Psychoanalytic Reading*. Baltimore, MA: Johns Hopkins University Press.

Munro, Lyle. 2012. "The Animal Rights Movement in Theory and Practice: A Review of the Sociological Literature: Animal Rights Movement in Theory and Practice." *Sociology Compass* 6 (2): 166–81. doi:10.1111/j.1751-9020.2011.00440.x.

Oatley, Keith. 2011. *Such Stuff as Dreams: The Psychology of Fiction*. Chichester, West Sussex; Malden, MA: Wiley-Blackwell.

Pashler, Harold, and Christine R. Harris. 2012. "Is the Replicability Crisis Overblown? Three Arguments Examined." *Perspectives on Psychological Science: A Journal of the Association for Psychological Science* 7 (6): 531–36. doi:10.1177/1745691612463401.

Pearson, Susan J. 2011. *The Rights of the Defenseless: Protecting Animals and Children in Gilded Age America*. Chicago, IL: University of Chicago Press.

Poe, Edgar Allan. 1994. "The Purloined Letter." In *Selected Tales*, 337–56. London: Penguin Classics.

Rorty, Richard. 2005. *Take Care of Freedom and Truth Will Take Care of Itself: Interviews with Richard Rorty*. Stanford, CA: Stanford University Press.

Rosenthal, Robert. 1979. "The File Drawer Problem and Tolerance for Null Results." *Psychological Bulletin* 86 (3): 638–41. doi:10.1037/0033-2909.86.3.638.

Rothgerber, Hank. 2014. "Horizontal Hostility among Non-Meat Eaters." *PLOS ONE* 9 (5): e96457. doi:10.1371/journal.pone.0096457.

Saunders, Marshall. 2015. *Beautiful Joe*. Edited by Keridiana Chez. Peterborough: Broadview Press.

Schlenker, Barry R. 1980. *Impression Management: The Self-Concept, Social Identity, and Interpersonal Relations*. Monterey, CA: Brooks/Cole Publishing Company.

Schmalhofer, F., and Charles A. Perfetti, eds. 2007. *Higher Level Language Processes in the Brain: Inference and Comprehension Processes.* Mahwah, NJ: Lawrence Erlbaum.

Serpell, J. A. 2004. "Factors Influencing Human Attitudes to Animals and Their Welfare." *Animal Welfare* 13 (1): 145–51.

Shaughnessy, John J, Eugene B Zechmeister, and Jeanne S Zechmeister. 2012. *Research Methods in Psychology.* New York: McGraw-Hill.

Shusterman, Richard. 1997. *Practicing Philosophy: Pragmatism and the Philosophical Life.* New York: Routledge.

———. 2000a. *Performing Live: Aesthetic Alternatives for the Ends of Art.* Ithaca, NY: Cornell University Press.

———. 2000b. *Pragmatist Aesthetics: Living Beauty, Rethinking Art.* 2nd ed. Lanham, MD: Rowman & Littlefield.

Sklar, Howard. 2013. *The Art of Sympathy in Fiction: Forms of Ethical and Emotional Persuasion.* Linguistic Approaches to Literature, volume 15. Amsterdam; Philadelphia, PA: John Benjamins Publishing Company.

Slovic, Paul. 2007. "'If I Look at the Mass I Will Never Act': Psychic Numbing and Genocide." *Judgment and Decision Making* 2 (2): 79–95.

Slovic, Scott, and Paul Slovic, eds. 2015. *Numbers and Nerves: Information, Emotion, and Meaning in a World of Data.* Corvallis, OR: Oregon State University Press.

Sorokowska, Agnieszka, Aleksandra Słowińska, Anita Zbieg, and Piotr Sorokowski. 2014. *Polska adaptacja testu Ten Item Personality Inventory (TIPI) – TIPI-PL – wersja standardowa i internetowa.* WrocLab. https://depot.ceon.pl/handle/123456789/5977.

Strange, Jeffrey J. 2002. "How Fictional Tales Wag Real-World Beliefs." In *Narrative Impact: Social and Cognitive Foundations,* edited by Melanie C. Green, Jeffrey J. Strange, and Timothy C. Brock, 262–86. Mahwah, NJ: L. Erlbaum Associates.

Taylor, Nik. 2016. "Suffering Is Not Enough: Media Depictions of Violence to Other Animals and Social Change." In *Critical Animal and Media Studies: Communication for Nonhuman Animal Advocacy,* edited by Núria Almiron, Matthew Cole, and Carrie P. Freeman, 42–55. Routledge Research in Cultural and Media Studies 77. New York: Routledge, Taylor & Francis Group.

Tedeschi, James T. 2013. *Impression Management Theory and Social Psychological Research.* New York: Academic Press.

Vezzali, Loris, Sofia Stathi, Dino Giovannini, Dora Capozza, and Elena Trifiletti. 2015. "The Greatest Magic of Harry Potter: Reducing Prejudice." *Journal of Applied Social Psychology* 45 (2): 105–21. doi:10.1111/jasp.12279.

Wales, Katie. 2011. *A Dictionary of Stylistics.* 3rd ed. Harlow, England; New York: Longman.

Walker, Alice. 2010. "Am I Blue?" In *Other Nations: Animals in Modern Literature,* edited by Tom Regan and Andrew Linzey, 182–87. Waco, TX: Baylor University Press.

Warrick, Jo. 2001. "'They Die Piece by Piece.'" *The Washington Post,* Accessed April 10, 2001. www.washingtonpost.com/archive/politics/2001/04/10/they-die-piece-by-piece/f172dd3c-0383-49f8-b6d8-347e04b68da1/?utm_term=.63950eaf0c71.

Webster, Murray, and Jane Sell, eds. 2007. *Laboratory Experiments in the Social Sciences.* Amsterdam; Boston, MA: Academic Press/Elsevier.

2 A Monkey, a Book, and Facebook, or How to Catch a Story in the Act

Provoking a Story

As any reader of crime fiction knows, sometimes the best way to prevent a crime is to provoke it. For instance, the police may be certain that a person engages in criminal activity, that the activity is extremely harmful to society, and that it therefore should be stopped as soon as possible. But as yet they may lack the evidence that would warrant an arrest.

In such cases, the police may decide to engineer a situation, a sting, that will lead the suspect to exhibit his or her typical criminal behavior or confirm his or her criminal intention. To this end, they might use agent provocateurs, detectives or confederates who will pretend that they want to engage in the criminal activity in which the suspect is involved. For instance, the agents will pretend that they want to join a terrorist network organized by the suspect, or buy from him or her a drug, or a bomb, or a colony of deadly bacteria. The agents will then not only have to play that role in a convincing manner but also manage the situation in such a way that it yields valid evidence and does not lead to any harm to third parties. This is, of course, not easy, as illustrated by those numerous stories in which the agents' cover is compromised and things go terribly wrong: people get hurt, property is destroyed, and the like. But despite the risks, there are times when the method of provocation seems to be the only or the best choice. And not only in detective work but also in psychological research. Here is why this was so in our case.

Just as it can happen to detectives, we too had a very elusive suspect. It was the animal narrative of a literary kind. We too wanted to catch it doing something we suspected it normally does (impacting attitudes), and we wanted to do so in its natural environment. This meant that we had to operate outside of laboratory spaces, in which psychological experiments are typically performed. Yet at the same time, we wanted to arrange the whole operation in such a way that it would yield valid evidence about the causal nature of that impact.

But just as sometimes happens to detectives, we too could not simply wait for our suspect to act. Practical considerations demanded that our project had to be completed in a limited time frame. Our only choice, then, was to provoke the suspect. To figure out how this could be done

DOI: 10.4324/9780429061424-3

was the very first task we set for ourselves when we embarked on our project. And we began by reminding ourselves what the natural environment of the suspect is and how the latter interacts with the key element of that environment, most importantly the readers.

So why do people read literary stories in the first place? Of course, some of them do so because they are told to (by a teacher or a parent, for instance), but given the examples of the successful literary animal narratives that inspired us, we were interested in so-called leisure reading. So again: why do people read stories when they read for themselves? And why do they choose some stories over others? For instance, what has made so many of them pick up a copy of Sewell's *Black Beauty* or W. Bruce Cameron's *A Dog's Purpose* or Sara Gruen's *Water for Elephants*?[1]

The first thing to note is that most people read literary stories that are also read by most people,[2] and that among the factors that spark their interest in such texts one of the most probable is a recommendation by somebody else, or the knowledge that others like them (Chevalier and Mayzlin 2006; Smith 2012; Phillips 2014, 80–84). There may be other factors that stimulate one's interest, but there is no doubt that the reading of literary stories for leisure is always driven by an interest, or curiosity (Smith 2009). It is also typically driven by the expectation of some kind of enjoyment: one hopes that a story will be pleasing in an aesthetic or an intellectual way (Wigfield 1997; Smith 2009).

So let us assume that you learn about a story which everybody is talking about, that your interest is stimulated, and that you expect it to be enjoyable. Then you get your hands on a copy. You go to a book store, or a library, or you download it to your electronic reader. And as soon as you are free from your duties (and sometimes you do not even wait for that), you find yourself a comfortable spot and begin to read. Of course, from then on, things may go in different directions. Your expectations may be frustrated or fulfilled. And it is trivially true that for narratives to do the job that is assigned to them by the moralists, the latter must be the case.

Everything was clear, then. What we needed for our provocation to work was a text that our subjects would be genuinely interested to read and one that they would be likely to find pleasing. A text that would additionally meet various formal and thematic requirements in order for us to be able to derive meaningful conclusions from the study. Of course, we also had to let the subjects read our story in a comfortable environment of their choice. And we needed a lot of subjects with certain demographic characteristics, not to mention that we needed to somehow divide them into two groups, a control and an experimental one.

In what was a moment that propelled our excitement for the whole project, it dawned upon us that possibly the best way to have such readers, such circumstances, and such a text is to engage a bestselling author as our agent provocateur. Bestselling authors tend to produce texts that are aesthetically pleasing to a lot of people, and because their texts have such qualities, a lot of people are interested in reading them. We would just have to convince such an author to write a story according to our suggestions, to publish it the way he or she normally publishes his or her works, and then take advantage of the natural interest of their readers in order to conduct our study.

Such a study would then have the qualities we needed. In particular, it would be possessed of a higher level of ecological validity than typical laboratory experiments. This is largely because it would "make use of a naturally occurring event" (Eysenck 2006, 282): in this case, the author publishing his output. The experiments which take advantage of such events are usually called "natural," and precisely because of the levels of ecological validity they allow they have been increasingly popular in social sciences in recent years. Let us now say a word on what they are about.

On Vapors, Sewers, and Cholera, or How to Perform a Natural Experiment

Although the popularity of natural experimentation is fairly recent, one of the *loci classici* of that method dates back to 1850s (Dunning 2012, 12). The experiment was performed in London by the physician John Snow and investigated the epidemiology of cholera. The dominant theory at the time held that cholera spread through air via miasmatic vapors, but the perceptive Dr. Snow saw that the predictive power of this theory was weak (Vinten-Johansen 2003, chap. 7 and 8). According to his favored hypothesis, the culprit was waste or waste-polluted water, and he saw one crucial piece of evidence for it in that "[d]uring London's cholera outbreak of 1853–54, … [the] addresses of deceased victims clustered around the Broad Street water pump in London's Soho district" (Dunning 2012, 13). For Snow, this suggested "that contaminated water supply from this pump contributed to the cholera outbreak" (13). But what this data showed was only a correlation, which, as we already know, cannot definitively indicate a causal connection. As we also know, the existence of such a connection can be established only through experiments. However, in this case, a typical laboratory experiment would have been unethical given the deadliness of the disease. A researcher could not have taken the risk of exposing his subjects to a disease such as cholera.

Yet, there then occurred an event over which Dr. Snow had no influence, nor which he could have prevented, but which at the same time could serve him as an experimental design. This is because at the time,

> [l]arge areas of London were served by two water companies, the Lambeth company and the Southwark & Vauxhall company. In 1852, the Lambeth company had moved its intake pipe further

upstream on the Thames, thereby 'obtaining a supply of water quite free from the sewage of London', while Southwark & Vauxhall left its intake pipe in place.

(Dunning 2012, 13)

This way, Dr. Snow could test his hypothesis on two populations of Londoners, who aside from one of them being exposed to waste-contaminated water (the "experimental" condition) were roughly the same in their make-up. What transpired was that the death rate in the "experimental" group was almost ten times higher than in the control group. The hypothesis was thereby confirmed experimentally in natural conditions. As Dr. Snow eloquently summarized his method:

> The mixing of the (water) supply is of the most intimate kind. The pipes of each Company go down all the streets, and into nearly all the courts and alleys. A few houses are supplied by one Company and a few by the other, according to the decision of the owner or occupier at that time when the Water Companies were in active competition. In many cases a single house has a supply different from that on either side. Each company supplies both rich and poor, both large houses and small; there is no difference either in the condition or occupation of the persons receiving the water of the different Companies ... It is obvious that no experiment could have been devised which would more thoroughly test the effect of water supply on the progress of cholera than this.
>
> (Snow 1965, 74–75, ct. after Dunning 2012, 13; cf. Snow 1857)

One difference between Snow's experiment (as well as most other experiments typically called natural) and the experiment we planned to perform would be this. While we wanted to make use of a particular, independently occurring, event, i.e. an author publishing his or her work, we also aimed to manipulate that event to an extent. It was as if Dr. Snow, knowing that the Lambeth Company had planned to relocate their intake pipes, had suggested to them that they move the pipes to a site above the sewage outlets. That would have been a specific intervention (or a provocation) on his part, and this is precisely the kind of strategy we intended to use. Now, the question was whether we could find a bestselling author who would be willing to work for us as an agent provocateur.

However improbable this may sound, it so happens that there was such an author. It was Marek Krajewski, one of the most popular Polish writers, whose work has been translated into more than 20 languages (including English, French, German, Spanish, Dutch, Swedish, Russian, Italian, and Hebrew), discussed in the media all around the world (including in top dailies such as *The Guardian* and *The New York Times*, (Brownell 2014)), and who also happened to have previously been an assistant professor at the faculty of philology at our university. Krajewski not only fit the bill, but also had one additional feature that made him the perfect author for our purposes. He is a writer of detective fiction, one of the most popular contemporary genres of literature (Faktorovich 2014, 1–2), so we could assume that his readership would be fairly representative of the reading public in general.

If the reader of this book thinks our luck was hard to believe, then we admit that the relationship between our idea of the natural experiment we needed and our choice of the writer might have been the reverse. Perhaps, this was our implicit memory that there exists a writer like that which prompted us to think about the experiment in question. We cannot be sure about this, as we are sure many researchers cannot be about the sources of their ideas.

How to Catch a Bestselling Author

An invitation e-mail was sent, then, to the author's official e-mail address. We were not even sure if we would get a reply. Eventually Krajewski's agent contacted us, asking on behalf of his client for more information about the project, and if Krajewski could meet with one of the research team. That latter task was assigned to the director of the project (Małecki). And while he would not consider himself someone who was easily impressed by writers and famous figures, given that a lot hinged on that conversation he was as stressed about it as a detective might be before an important interrogation.

Fortunately, the author agreed, revealing, to Professor Małecki's surprise, that he was an animal lover himself, sharing his life with a dog who he had taken from a shelter after being saved from abusive owners. Not only this, Krajewski was an aficionado of evolutionary psychology, including the work of another member of the research group (Pawłowski), and the fact that the latter was on our team was the main reason why he decided to join us (Ślązak 2017). Once he was on board, we engaged in a series of meetings during which we discussed the details of his writing task, something which added a unique quality to our research.

The text we commissioned from him was to be included in his forthcoming book, *The Lord of the Numbers* (2014), and he used our meetings to test his ideas for that novel, and to ask us for an expert's advice. For instance, he wanted to know about how an organism reacts to being burned with cigarettes, which is something Pawłowski, the biologist among us, was able to explain in detail. So here we were able to watch a writer doing research, and we must admit that our belonging to a very narrow group of people who knew well in advance what dozens of thousands of readers would love to know, and that we were of course officially obliged to keep all this secret, added an extra layer of excitement to our work.

Detectives and Other Animals

One interesting thing about our cooperation is that at no point did the author mention that including an animal story in his forthcoming book would be something even remotely difficult. Of course, this might have had to do with the fact that Krajewski is a professional author who writes at least one book per year, and such authors often are sufficiently skilled and productive to write about almost any subject. There is simply no place here for being overwhelmed by creative challenges, for writer's block, and the like. But one additional thing that certainly helped is that detective fiction from its very beginnings has often assigned key roles to animals (Herbert 2003). They have often occupied the roles of instruments, accomplices, or even criminals themselves in such stories. They have been witnesses, detectives' assistants, and worked as detectives as well (McHugh 2011, 27–64). And of course, they have been victims.

It would be impossible here to provide even a brief history of such motifs, but by way of illustration, recall, for instance, that among the classics of detective fiction penned by Edgar Allan Poe, there is not only "The Purloined Letter," which we talked about in Chapter 1, but also "The Murders in the Rue Morgue" (Poe 1994, 118–53; Boggs 2013, 109–32; Peterson 2013, 22–49). In that story, the mysterious killer of two women, whom the witnesses hear speaking an unidentifiable language, turns out to be an exceptionally clever domesticated orangutan. The case is eventually solved by the famous C. August Dupin himself (the hero of the mystery of the purloined letter), but other famous literary detectives had to deal with animals too.

We could mention here Sherlock Holmes in Sir Arthur Conan Doyle's *The Hound of the Baskervilles* or the FBI agent Clarice Starling in Thomas Harris's *The Silence of the Lambs* and *Hannibal*, who was immortalized by Jodie Foster and Julianne Moore in the famous film adaptations of these two books (Harris 1999a, 1999b; Tasker 2009; Doyle 2012). Some of the most revolting and uncanny elements of those incredibly revolting and uncanny novels (and incredibly revolting and uncanny movies based upon them) prominently involved other species, including a sphinx moth lodged in the throat of a severed human head.[3]

Given that animals have often featured in the thematic repository of detective fiction, it is no wonder that they were to be found in Krajewski's novels even before our cooperation, beginning with his first book *Death in Breslau* (2013). That novel, which catapulted him to stardom in Poland when it was first published in 1999, had the working-title of *Scorpions* due to the significant role played by those arachnids in the story. The role assigned to them is also quite scary, as it involves swarming in the intestines of victims of brutal murders, not to mention acting as the symbol of a morbid cult responsible for the crimes in question.

But we did not want a story in which an animal would be an instrument or a symbol of the crime. Neither did we need animal detectives or animal murderers. Consistent with our investigative assumptions, we wanted to have an animal victim. The novel in which it was to feature had already been partly sketched and involved a murderous mathematician who killed his victims according to a complicated pattern. Our animal was to be one of his prey.

But then we had to decide which species it would belong to. Given that the novel was set in mid-twentieth century Poland, the easiest choice for the writer was a species that was common here at the time, a pet such as a dog or a cat, or a farm animal such as a pig, a cow, or a hen. But in this particular experiment, we wanted to avoid a species whose welfare is typically respected or disregarded in the West, with the dog and the cat belonging to the former category and most farm animals belonging to the latter. We also wanted an animal who would be perceived as sufficiently close or similar to humans in order for its capacity for suffering be widely acknowledged by the readers. Eventually, we settled on a monkey (Mitchell, Thompson, and Miles 1997; Ayala and Cela-Conde 2017).

We also did not want the monkey's suffering to be merely the fault of the psychopath's. If it were, it might have been perceived by the readers as detached from the widespread systematic suffering of animals that we were most concerned with, and which was indexed by our scale (Herzog, Betchart, and Pittman 1991; Herzog and Mathews 1997). Finally, we wanted the monkey to be a fully rounded character to whom readers could become attached, similarly to the way they can become attached to the animal protagonists of *Black Beauty* or *Babe* or John Steinbeck's *The Red Pony* (1994). We believed that this demanded presenting a believable life story of the animal (Tsovel 2005), from its early years to its ordeal at the hands of the psychopath. This, then, was our wish list that we presented to the writer and which we discussed with him during our meetings. As soon as Krajewski decided he had a clear grasp of what we wanted, he told us he would begin working on the manuscript and contact us as soon as his work was done, and within our deadline of three months to start the experimental phase of our project.

We did not hear from him at all during that period, and it turns out that when he works on a manuscript, Krajewski becomes a virtual recluse, limiting his other activities to a bare minimum. But when the agreed submission date came round, we received by e-mail the animal story that was to be included in the novel. It was just as we wanted it to be, and even more (see Appendix 2).

The text spun a narrative of the monkey's life. It spanned across 15 years, and included elements that might help the reader form an emotional attachment to the animal, beginning with her rearing in a Venezuelan jungle and her close relationship with her mother. And once the stage was set in this way, the story presented a series of misfortunes to which humans submitted the animal, a series that was at the same time gripping, believable, and historically accurate (as a philologist,

Krajewski carefully researches his novels, almost to the point of pedantry). The story, in summation, went as follows:

The monkey and her mother are first captured by natives of the country where they live in the wild. They are separated and each sold to animal traders. Our protagonist eventually finds herself in Amsterdam, to which she is taken on board of a Dutch freighter. Sold to a travelling circus, she has to endure a genuine horror at the hands of a sadistic trainer, who burns her with cigarette butts when conditioning her to perform tricks:

> The penetrating shriek of the animal spanned a few registers. The monkey's body, wrapped in an enormous hood, trembled spasmodically, and her nervous system reacted by relaxing sphincters.
>
> The trainer withdrew his hand in disgust, grabbed the creature by its hand, dragged the tiny, still trembling body through the sand of the arena, and then, having waited for an hour, he would scrape off the dust covered muck.
>
> (cf. Krajewski 2014)

Once the circus owners consider her useless, she is sold again, this time to a Polish

> organ grinder, who did not want anything from her apart from sitting on his organ. He looked after her with such care that he did not even economize on lamp oil and put a lit lamp next to her cage at night.

Unfortunately, following the organ grinder's death, the monkey then finds herself in the hands of the psychopath, the titular Lord of the Numbers.

He names her Clotho "as a tribute to one of the mythical weavers, who – together with her sisters Lachesis and Atropos – weaved the thread of human existence." But it is he who is the weaver of her existence, which involves submitting her to a series of quasi-scientific experiments that are cryptically related to his murders of humans in the novel:

> The man would put inside the cage an iron stand with two ladders leading to a small platform. One of them was black, the other white. Lying on the platform, there was a walnut. The creature would happily climb for the walnut – using either the white or the black ladder. Then the man would draw out two protruding wires in her direction. Electricity would twist her body and force a high-pitched shriek out of the tiny throat. The man would smile friendly, say something in a silent voice and touch one or the other ladder with a pointer – the white and the black one, in turns. Clotho did not know what was

on her tormentor's mind. Afraid of the wires, she jumped from one ladder to the other like crazy, blindly. Then the man would apply electric shocks again. Apparently, he demanded something else. She did not understand that he wanted to make her disorderly jumps less chaotic – that all that he wanted was that she first climbed the black ladder, and then immediately the white one.

Clotho failed to grasp the man's intentions. She was helpless. All she could do was to look into the eyes of the tormentor approaching her. And then to suffer.

(cf. Krajewski 2014)

Krajewski had provided more than we expected: all the ingredients we wanted the author to include were not only there but had been turned into a gripping whole, with the use of devices that we neither alluded to nor even thought about, but which we perhaps could have expected from such a fine literary craftsman. For instance, the power of literary storytelling lies, among other things, in seemingly peripheral details (Barthes 2010), and the story contained many, beginning with the fact that the author specified that our generally defined "monkey" would be a capuchin.

Another, and related, feature of a good literary story, is that it always contains an ethical gray area: the ambiguities and complexities that make definite moral judgments difficult and which separate it from polemic or moralizing hack-jobs (Foulkes 1983; Hakemulder 2000, 23–24; cf. Nussbaum 1990; Johnson 2014). Our author himself was particularly sensitive to ambiguities and gray areas. This is because the kind of hard-boiled detective fiction he specializes in resides in such territories, which is nicely captured by an adjective it is usually described with – "noir." The frisson that is generated by noir detective fiction lies not simply in the gruesome details of the crimes it deals with. It also lies, and importantly so, in that the line between the detective and the criminal is often barely distinguishable: even positive protagonists tend to have a dark side to them, and, perhaps even more frighteningly (Moore 2006), there is often something redeeming about even about the worst antagonists, which complicates our judgment of them.

A story about an idealized animal victim and her demonized human oppressors would smack too much of simple-minded propaganda to be convincing as a literary text. So the author made Clotho bite an innocent human child at some point of the story, and he additionally made that child into one of the human characters with a positive attitude toward our animal. Another such character was the organ grinder, who not only nuanced the picture of human attitudes toward animals in the story but also added to the dynamic of the plot. A uniform string of misfortunes would be hardly believable, and eventually numbing. The short idyll which the capuchin experienced with the man helped to avoid that and to make the torments induced by the Lord of the Numbers even more shocking, revolting, and touching. So here we had a story that was *both* specifically designed for our purposes and possessed genuine literary qualities. This is how we made use of it.

A Monkey, a Book, and Facebook: Our Natural Experiment

Four weeks before the official publication date (Sept. 11, 2014) of the *Lord of the Numbers*, the author announced on his public Polish Facebook profile a quiz that served as a cover for our study. It was also announced on his publisher's Facebook profile and on the website of a popular Polish book lovers' community. The quiz offered an opportunity to read a fragment of Krajewski's then still unpublished novel. It also offered the opportunity to win a free copy of the book in exchange for answering a set of questions. As Krajewski explained in his Facebook post: "I am currently cooperating with scholars who would like to study the psychological profile of the readers of my novels by using an internet questionnaire" (see Appendix 1).

We are happy to say that this kind of cover worked even better than we had planned. Not only did none of Krajewski's readers expressed any suspicions on his Facebook profile but many of them got apparently excited about the idea and even speculated on what kind of use the author might make of such psychological data. Some of the speculations were quite entertaining, e.g., the one that Krajewski may use this material as an inspiration for his future fictional characters.

Once they accessed a special website and agreed to the conditions of the quiz, the participants were randomly assigned to one or the other of the following two experimental conditions: the opportunity to read online a three-page fragment of the novel (see Appendix 2) that concerned the plight of Clotho (the experimental group), or alternatively, the opportunity to read a fragment of similar length (see Appendix 3) in which the main protagonist of the novel, the private detective Edward Popielski (the protagonist of a popular series of Krajewski's novels) is approached by a stranger with a request to solve an as yet unspecified case – a subject we had deemed neutral from the perspective of our study (the control group).

Immediately after reading the text, the subjects filled out an online questionnaire whose ostensible purpose was to examine the psychological profile and worldview of Krajewski's readers and their impressions about the text they read (see Appendix 4). The questionnaire consisted of 53 items scored on a 7-point scale, where 1 meant "I completely disagree" and 7 meant "I completely agree." Camouflaged among items concerning personality traits as well as moral and political beliefs were the items constituting our Attitudes Toward Animal Welfare scale (hereafter ATAW scale), which measured our subjects' attitudes toward animals. At the end of the questionnaire, participants provided demographical data, including whether the subjects keep pets.

A professional market research agency was hired to create and manage the online questionnaire, as well as to design a special website through which the questionnaire was accessed by the subjects. To minimize the risk of a given person participating in the study more than once, two measures were used: http cookies that blocked access to the questionnaire from the same web browser once it had been filled out; and the verification of the personal data that the participants submitted in order to take part in the quiz.

To reduce the possibility of communication between the participants interfering with the results, they were asked to abstain, for the duration of the quiz, from revealing any details about the questionnaire or the texts they had read, including on the author's Facebook profile. For the same reason, while the participants were given a chance to opt out of the study after completing the survey, they were not debriefed. To our best knowledge, no relevant details were revealed publicly and no participant expressed suspicion as to its real purpose.

Whodunit (to the Readers)? or Our Results

Within the period of 19 days for which the experiment was running, the quiz attracted 1833 Polish readers, 89% of whom participated on the first three days. Among them, there were 1241 women, aged between 14 and 81, and 592 men, aged between 15 and 69. This was an impressive sample, both in terms of its size and the demographic diversity. We had never heard of any experimental study on literary reading that would involve so many people. Clearly, Krajewski's stardom could work miracles. Now we only had to see what this data showed.

To verify whether our experimental setting influenced attitudes toward animal welfare, we performed our statistical analyses with so-called pairwise comparisons. To be more exact, we wanted to see not only if there is a difference in attitudes toward animal welfare between the participants from the control and the experimental group, but also between women and men in our sample, and between those who had a pet and those who did not. Our analyses revealed that women expressed more pro-animal welfare attitudes than men ($p < 0.0001$, $\eta^2 = 0.09$), and there was also a significant main effect of pet possession ($p < 0.0001$, $\eta^2 = 0.07$), indicating that participants who declared having a pet at home scored higher in ATAW as compared to those who did not report possessing a pet. But, most importantly, we also found a significant main effect of our experimental condition ($p < 0.00001$, $\eta^2 = 0.02$), indicating that the participants from the experimental group (who read the text about the abused monkey) scored higher in ATAW than participants from the control group (Figure 2.1).[4]

Figure 2.1 The influence of the experimental conditions on attitudes toward animal welfare.

The hypothesis was confirmed: our experiment showed that the literary narrative used in our experiment influenced the subjects' attitudes toward animal welfare in the sense of making their attitudes more pro-animal welfare. Since this effect was observed in almost all tested groups, even in the group that presented the least pro-animal welfare attitudes (i.e. men who did not own pets), our data also confirm that the effect was not due to a specific sample that is sensitive to the well-being of animals in general! In other words, animal stories can affect various kinds of readers.

But as we said earlier in the book, results are almost always more interesting than they appear on the surface. Consider, for instance, that our data corroborate the results of experimental studies showing that narratives depicting the plight of an individual member of a given group (e.g. a drug addict) can help improve attitudes toward that group as a whole (Batson et al. 1997; Green and Brock 2000; Johnson et al. 2013). One significant difference between this research and ours is that the group comprising all animals is a much larger and varied out-group than any human out-group could possibly be (Plous 2003). It is precisely for this reason that "the animal," understood as a category hierarchically opposed to "the human," has been called "monstrous." As Kelly Oliver points out, that category "erases vastly diverse differences among individual animals and subgroups within species and between species themselves. ... it herds countless species into one category and then denigrates them" (2009, 34; cf. Derrida 2009). That a representation of one of its representatives had an effect on attitudes toward all of them generally conceived is striking indeed.

What makes our data even more striking is that the items with which we measured the participants' attitudes included some that concerned not animals in general, but particular species that were unrelated to the one included in the narrative. They concerned, too, issues that were not directly related to the topic of the narrative. While it seems prima facie understandable why the story we used might make one express less agreement with the statement such as "The suffering of animals is an acceptable price for inventing drugs for humans," it is less understandable why it would make one change his or her mind whether "The slaughter of whales and dolphins should be immediately stopped even if it means that some people will be put out of work," as it in fact did in our case. Clearly, there is no direct logical link here, something which supports the view that how our "cognitive/ethical norms and intuitions [about other species] are formed, stabilized, and transformed" is not necessarily a matter of ethical deliberation (Herrnstein Smith 2012, 156), and that, more generally, empathetic feelings have a significant yet complex impact on our moral judgments (Batson et al. 1995; Batson et al. 1997; Prinz 2011; Batson 2015; Singer 2016).

Our results become even more significant if we recall how the experiment approached the question of ecological validity. To reiterate, we used a genuine literary text and performed our study outside of laboratory settings. It may be assumed, too, that our subjects were genuinely interested in reading the text and following its plot, and that they would have been inclined to read it independently of the experimental conditions. In addition, it is now an established practice that authors publish online sample material from their forthcoming books for marketing purposes, so the fact that the participants read only a fragment of the novel and that they did so from their computer or tablet screens most likely did not seem extraordinary to them. The design of our study made it practically impossible for the subjects to guess its purpose, therefore minimizing the risk of impression-management occurring. It did seem, then, that we caught our suspect red-handed in its natural environment. An animal story really can improve attitudes toward animals.

But note, finally, that there is something significant about our provocation even apart from its result. For in conducting our investigation, we helped to create the suspect ourselves. We gave life to a certain hybrid: a literary-scholarly story composed of parts that fit our research purposes. And just like the most famous literary creator of a hybrid life form, Dr. Frankenstein, we could not fully control our creation. The story became a part of a bestselling novel. It was out there in the world, living a life of its own. We would not be able to stop it from affecting thousands of other people, even if we wanted. But we of course did not want to. Fortunately, we had just proven that the deeds of our creation would be beneficial.

Notes

1 These are two recent examples of very successful animal-themed novels (Gruen 2006; Cameron 2010). They are both bestsellers and both were adapted for screen. For a discussion of *Water for Elephants* from the perspective of animal studies, see Caesar (2009, 126–28).
2 Of course, anecdotal evidence tells us that there are also people who try to read only those stories that are not read by most people, preferably by nobody else. But everyday experience tells us also that they are a minority, something which they will most likely be happy to hear, as they apparently act this way mostly in order to seem special.
3 It is worth noting here that one of the foundational texts of animal studies is Cary Wolfe's paper on Jonathan Demme's movie *The Silence of the Lambs*, first published in 1995 in the journal *Boundary 2* and later included, as chapter 3, in Wolfe (2010).
4 In order to address the potential worry that the inclusion in our questionnaire of an item concerning apes might have skewed the results of our study (which used a text about a monkey) we have performed additional analyses which excluded that particular item. The general results were the same as reported above.

Works cited

Ayala, Francisco José, and Camilo J. Cela-Conde. 2017. *Processes in Human Evolution: The Journey from Early Hominins to Neandertals and Modern Humans*. Oxford; New York: Oxford University Press.

Barthes, Roland. 2010. "The Reality Effect." In *The Rustle of Language*, translated by Richard Howard, 141–48. Berkeley: University of California Press.

Batson, C. Daniel. 2015. *What's Wrong with Morality? A Social-Psychological Perspective*. New York: Oxford University Press.

Batson, C. Daniel, Eddie Harmon-Jones, Heidi J. Imhoff, Erin C. Mitchener, Lori L. Bednar, Tricia R. Klein, Lori Highberger, and Marina Polycarpou. 1997. "Empathy and Attitudes: Can Feeling for a Member of a Stigmatized Group Improve Feelings toward the Group?" *Journal of Personality and Social Psychology* 72 (1): 105–18. doi:10.1037/0022–3514.72.1.105.

Batson, C. Daniel, Tricia R. Klein, Lori Highberger, and Laura L. Shaw. 1995. "Immorality from Empathy-Induced Altruism: When Compassion and Justice Conflict." *Journal of Personality and Social Psychology* 68 (6): 1042–54.

Boggs, Colleen Glenney. 2013. *Animalia Americana: Animal Representations and Biopolitical Subjectivity*. Critical Perspectives on Animals: Theory, Culture, Science, and Law. New York: Columbia University Press.

Brownell, Ginanne. 2014. "Move Over Scandinavian Noir, Here Comes the Polish Gumshoe." *The New York Times*, March 19, 2014. www.nytimes.com/2014/03/20/arts/international/move-over-scandinavian-noir-here-comes-the-polish-gumshoe.html.

Caesar, Terry. 2009. *Speaking of Animals: Essays on Dogs and Others*. Human-Animal Studies, v. 7. Leiden; Boston, MA: Brill.

Cameron, W. Bruce. 2010. *A Dog's Purpose*. 1st ed. New York: Forge.

Chevalier, Judith A, and Dina Mayzlin. 2006. "The Effect of Word of Mouth on Sales: Online Book Reviews." *Journal of Marketing Research* 43 (3): 345–54. doi:10.1509/jmkr.43.3.345.

Derrida, Jacques. 2009. *The Animal That Therefore I Am*. Translated by David Wills. New York City: Fordham University Press.

Doyle, Arthur Conan. 2012. *The Hound of the Baskervilles*. London: Penguin.

Dunning, Thad. 2012. *Natural Experiments in the Social Sciences: A Design-Based Approach*. Cambridge; New York: Cambridge University Press.

Eysenck, Michael W. 2006. *Psychology for AS Level*. Hove, East Sussex: Psychology Press.

Faktorovich, Anna. 2014. *The Formulas of Popular Fiction: Elements of Fantasy, Science Fiction, Romance, Religious and Mystery Novels*. Jefferson, NC: McFarland & Company, Inc., Publishers.

Foulkes, A. Peter. 1983. *Literature and Propaganda*. New Accents. London; New York: Methuen.

Green, Melanie C., and Timothy C. Brock. 2000. "The Role of Transportation in the Persuasiveness of Public Narratives." *Journal of Personality and Social Psychology* 79 (5): 701–21.

Gruen, Sara. 2006. *Water for Elephants: A Novel*. 1st ed. Chapel Hill, NC: Algonquin Books.

Hakemulder, Jèmeljan. 2000. *The Moral Laboratory: Experiments Examining the Effects of Reading Literature on Social Perception and Moral Self-Concept*. Utrecht Publications in General and Comparative Literature, v. 34. Amsterdam; Philadelphia, PA: J. Benjamins Pub.

Harris, Thomas. 1999a. *Hannibal*. London: Orbit.

———. 1999b. *The Silence of the Lambs*. New York: St. Martin's Paperbacks.

Herbert, Rosemary. 2003. "Animals." *Whodunit? A Who's Who in Crime & Mystery Writing*. New York: Oxford University Press.

Herrnstein Smith, Barbara. 2012. *Scandalous Knowledge: Science Truth and the Human*. Cambridge: Cambridge University Press.

Herzog, Harold A., and Steve Mathews. 1997. "Personality and Attitudes toward the Treatment of Animals." *Society and Animals* 5 (2): 169–75.

Herzog, Harold, Nancy S. Betchart, and Robert B. Pittman. 1991. "Gender, Sex Role Orientation, and Attitudes toward Animals." *Anthrozoos: A Multidisciplinary Journal of the Interactions of People & Animals* 4 (3): 184–91. doi:10.2752/089279391787057170.

Johnson, Dan R., Daniel M. Jasper, Sallie Griffin, and Brandie L. Huffman. 2013. "Reading Narrative Fiction Reduces Arab-Muslim Prejudice and Offers a Safe Haven From Intergroup Anxiety." *Social Cognition* 31 (5): 578–98. doi:10.1521/soco.2013.31.5.578.

Johnson, Peter. 2014. *Moral Philosophers and the Novel: A Study of Winch, Nussbaum and Rorty*. Houndmills, Basingstoke; New York: Palgrave Macmillan.

Krajewski, Marek. 2013. *Death in Breslau*. Translated by Danusia Stok. New York: Melville House.

———. 2014. *Władca liczb*. Kraków: Znak.

McHugh, Susan. 2011. *Animal Stories: Narrating Across Species Lines*. Posthumanities, v. 15. Minneapolis: University of Minnesota Press.

Mitchell, Robert W., Nicholas S. Thompson, and H. Lyn Miles. 1997. *Anthropomorphism, Anecdotes, and Animals*. Albany: State University of New York Press.

Moore, Lewis D. 2006. *Cracking the Hard-Boiled Detective: A Critical History from the 1920s to the Present*. Jefferson, NC: McFarland & Co.

Nussbaum, Martha Craven. 1990. *Love's Knowledge: Essays on Philosophy and Literature*. New York: Oxford University Press.

Oliver, Kelly. 2009. *Animal Lessons: How They Teach Us To Be Human*. New York: Columbia University Press.

Peterson, Christopher. 2013. *Bestial Traces: Race, Sexuality, Animality*. 1st ed. New York: Fordham University Press.

Phillips, Angus. 2014. *Turning the Page: The Evolution of the Book*. London; New York: Routledge.

Plous, Scott. 2003. "Is There Such a Thing as Prejudice Toward Animals?" In *Understanding Prejudice and Discrimination*, edited by Scott Plous, 509–28. Boston, MA: McGraw-Hill.

Poe, Edgar Allan. 1994. "The Murders in the Rue Morgue." In *Selected Tales*, 118–53. London: Penguin Classics.

Prinz, Jesse. 2011. "Is Empathy Necessary for Morality?" In *Empathy: Philosophical and Psychological Perspectives*, edited by Amy Coplan and Peter Goldie, 211–29. Oxford; New York: Oxford University Press.

Singer, Peter. 2016. "The Empathy Trap." *Project Syndicate*, December 12, 2016. www.project-syndicate.org/commentary/danger-of-empathy-versus-reason-by-peter-singer-2016-12.

Ślązak, Anna. 2017. "To Nie Truizm. Czytanie Naprawdę Zmienia Świat." *Nauka w Polsce*. January 12, 2017.

Smith, Kelvin. 2012. *The Publishing Business: From p-Books to e-Books*. Lausanne; La Vergne, TN: AVA Academia; Distributed in the USA by Ingram Publisher Services Inc.

Smith, M. Cecil. 2009. "Literacy in Adulthood." Edited by M. Cecil Smith and Nancy DeFrates-Densch. *Handbook of Research on Adult Learning and Development*. New York: Routledge.

Snow, John. 1857. "Cholera, and the Water Supply in the South Districts of London." *British Medical Journal* 1 (42): 864–65.

———. 1965. *Snow on Cholera*. New York: Commonwealth Fund, Oxford University Press.

Steinbeck, John. 1994. *The Red Pony*. Penguin Twentieth-Century Classics. New York: Penguin Books.

Tasker, Yvonne. 2009. *The Silence of the Lambs*. Basingstoke: Palgrave Macmillan.

Tsovel, Ariel. 2005. "What Can a Farm Animal Biography Accomplish? The Case of Portrait of a Burger as a Young Calf." *Society and Animals* 13 (3): 245–62.

Vinten-Johansen, Peter, ed. 2003. *Cholera, Chloroform, and the Science of Medicine: A Life of John Snow*. Oxford; New York: Oxford University Press.

Wigfield, Allan. 1997. "Reading Motivation: A Domain-Specific Approach to Motivation." *Educational Psychologist* 32 (2): 59–68. doi:10.1207/s15326985ep3202_1.

Wolfe, Cary. 2010. *What Is Posthumanism?* Posthumanities Series, v. 8. Minneapolis: University of Minnesota Press.

3 Does It Matter If It Is True?

On Slaughterhouses, Fiction, and Non-Fiction

Interrogating a Story

If natural experiments are like using agents provocateurs, then laboratory studies would be the scientific equivalent of yet another crucial detective technique, i.e., interrogation. And just as interrogation has its specific advantages as an investigatory technique when compared with the use of agents provocateurs, so do laboratory experiments have advantages over natural experiments. Consider, for instance, that agent provocateurs are drastically constrained in their actions by the role they play. Whatever they do, this must not look like something the police would do. Consider too, that this kind of investigation often has to take place in the suspect's natural environment, one that he or she usually knows better than the detectives and one that they cannot entirely control. This gives the suspect a definite advantage in evading detection.

In an interrogation room, in contrast, detectives do not have to pretend to be someone they are not, and it is also an environment on which they exert almost complete control. The suspect cannot get out of the interrogation room or hide from view. And no less importantly, that particular environment can be carefully arranged by the detectives so that they can play whatever games they think they need (and are permitted by law) to manipulate the suspect in order to obtain the sought-after information. The suspect is served up on a plate, so to speak, and this can make the investigation easier. For instance, whereas it might be difficult or impossible without arousing suspicion for undercover detectives to ask their suspect questions about certain details relevant to the investigation, in the interrogation room they can ask about such things as openly as they please.

Laboratories are psychologists' interrogation rooms, where they have suspected phenomena on their plate and can manipulate them in full view and in the minutest details (Shaughnessy, Zechmeister, and Zechmeister 2012, chap. 4, 6, 10; cf. Webster and Sell 2007). In the series of laboratory experiments which we describe in this and the subsequent two chapters, we had tried to take advantage of this. We wanted to see whether animal narratives tend to do what we suspected them of doing.

DOI: 10.4324/9780429061424-4

And further, we wanted to understand the mechanisms by which they do it. For instance, it is widely agreed that the psychological impact of stories differs depending on whether they are perceived as fictional or not, and there have been numerous theoretical and empirical studies devoted to that topic (Green and Brock 2000; Batson et al. 2002; Strange 2002; Mar et al. 2006). This was important to us because among the most famous examples of the attitudinal impact of animal stories we find both works of fiction, such as *Black Beauty,* and journalistic accounts, such as "They Die Piece By Piece." It was only natural for us, then, to devote one of our laboratory experiments to seeing whether *animal* narratives perceived as fictional would have a different impact on attitudes toward animal welfare than those that are perceived as non-fictional. Its results are related in this chapter. Other features of texts on which their attitudinal impact might depend and that we chose to study in our laboratory experiments include whether they contain evaluative hints and arguments, whether they are written in first- or third-person voice, and which species their animal protagonist belongs to. The results of these studies are related in the next two chapters.

So this is what our laboratory experiments were about. Now let us say a word on the methods we used, or how we interrogated our narrative suspects. The first thing that needs to be noted is that our laboratory experiments were not performed in *actual* laboratories. They were performed instead in improvised laboratory spaces: classrooms and lecture halls of the institutions attended by our subjects, high school and university students. One advantage of this was that these spaces did not feel entirely alien or discomforting to the subjects. Of course, there is no denying that sometimes a classroom may seem to a student to be as alienating or discomforting as a laboratory would be (or a police interrogation room for that matter), but we think that normally, it would feel less alienating than an actual psychological laboratory with its sterile cubicles.

No less importantly, and again unlike psychological laboratories, classrooms were naturally recognized by our subjects as some kind of *reading* environment, as this is where they had read texts on a regular basis. This was definitely an advantage from the perspective of research such as ours that is concerned with reading. Aside from that, these spaces had most of the defining features of a laboratory. The most important of those was their confined character. This allowed us to observe the subjects at all times and be sure that we controlled the potential interfering factors or at least were aware of them; that, for instance, the subjects in one of the groups were not watching a TV documentary on animal abuse during our experiment, and the like.

The basic procedure was the same in all our experiments and in order to avoid tiresome repetitions throughout this and the following chapters, let us present its general description here. In the first phase, the subjects were randomly assigned to however many groups we needed for a given study. In each study, we would always have one control group and some experimental groups whose number ranged from two in some studies to as many as nine groups, as in the study on the species of the protagonist, where we varied our experimental text between nine species of its protagonist, from a chimpanzee to a lizard.

In the second phase, each group would be taken to a separate room, where they would be informed by our confederates, a man and a woman, about the ostensible purpose of our study, i.e., to study the relations between the psychological profile of readers and their perception of texts. The confederates would then administer a narrative and wait until all the subjects have finished reading it. At the next stage, they would explain to the subjects that they were to complete a questionnaire, and give them instructions for doing so. Then, they would distribute the questionnaires and collect them when everyone had completed them.

In general, the questionnaires we used in our studies looked similar to the one we used in the experiment described in Chapter 2 (see Appendix 4). They would comprise of a part devoted to the personality and worldviews of the participants (including the ATAW scale), a part which concerned the text itself, and a set of demographic questions. While they were roughly the same in their make-up, some of them included items that were specific to a given study. We will describe each such case in detail in the pages that follow.

Note from that description of our general method that we had both a man and a woman as our confederates directing the subjects as they participated in the experiments. This is not a trivial detail. Another thing that decades of psychological research have taught psychologists is that the gender of the experimenter matters. It matters because of so-called "experimenter effects," that is, the various ways in which what the experimenter is like, or what he or she does, can influence the results of the study (Webster and Sell 2007, chap. 6; Hendrick and Jones 2013, 62–67). In particular, it is now common knowledge that the response of women and men to an experimental task, including filling out questionnaires that measure attitudes, may vary depending on whether the experimenter is of the opposite gender or not (Harris 1971; Argentino, Kidd, and Bogart 1977; Williams et al. 1993; Nichols and Maner 2008). Since we knew that we would have groups consisting of both male and female students and that sometimes their proportions would be uneven, we wanted to prevent this factor from distorting our results. So we decided to have both a man and a woman as confederates.

But why did we use confederates in the first place? Wouldn't it have been easier had we worked with the subjects directly? After all, no one could know the design of our experiments better than we did, and therefore nobody could know better how to perform them correctly, e.g., which potentially distorting phenomena to pay attention to during the running of the studies, and the like. Also, there was no one who would have been more engaged in the project and therefore nobody more determined to invest the proper time and effort to ensure it was conducted properly.

Well, this was precisely the problem. We were *too* engaged in the project to work with the subjects ourselves. Recall our vitamin C example and how we mentioned that a researcher may inadvertently suggest to the subjects, through gestures, verbal allusions, and other kinds of hints what the expected result is (Shaughnessy, Zechmeister, and Zechmeister 2012, 200–201). It is reasonable to assume that the stronger the researcher's engagement, or emotional involvement in the project, the larger the probability that this kind of effect may occur. So, paradoxically, the more we wanted to be there to make sure the experiments would proceed as planned, the more important it was that we kept away in order to achieve this.

How to Recognize Fiction When You See It?

Our first laboratory experiment focused on the difference in attitudinal impact between texts perceived as fictional and those perceived as non-fictional. Admittedly, the phrase "perceived as fictional or non-fictional" may sound unnecessarily pedantic. Shouldn't we rather say, simply, "were fictional" or "were non-fictional"? Things are *not* so simple. It so happens that the actual fictionality or otherwise of the text does not *necessarily* manifest itself in its formal features. There are texts which, in themselves, without any additional information about their genre or the intentions of the author, may be equally well perceived as fictional and as non-fictional. Therefore, a text that is meant by the author as a factual account may be perceived by her readers as fiction, and vice versa.

Consider, for instance, the following fragment from Stephen S. Brandom's detective novel *Darkness Over Chicago* (1999). The main protagonist of the novel, detective John McDowell, is conducting an investigation about the ties between the New York and Chicago Mafias and various industries, including the art market, gastronomy, and food production. In the following extract, the detective arrives at a state prison to meet with an inmate who was involved in the illegal meat trade:

> Steve Parrish finally entered and was directed to where I was seated.
> Parrish, a compact, graceful African American in his early thirties, crossed the room with a slow, confident gait. He sat down across from me, smiled, then asked – as if he were actually interested – how I was doing. I wanted to establish a good rapport, so we talked for a while about life in general, his specifically.

He had worked in slaughterhouses ever since he was a teenager. He didn't say much about his life in Chicago's streets, the dangerous crowd he ran with – or how he'd landed in jail this time. I asked him what kind of horses were slaughtered at the plants where he'd worked.

"Belgians, Arabians, little ponies- all kinds. Long as it's a horse. Stolen horses, too."

"Stolen?" I asked. Lately at my job, I'd been getting a lot of complaints about horses that had been stolen from their owners and sold to slaughterhouses. Horse theft for slaughter seemed to be on the increase (Brandom 1999).

If you didn't know the source of the extract, wouldn't it be possible to believe that this is a fragment of a journalistic report on the dark underbelly of the meat industry? After all, it seems like something that could come, for instance, from a book such as Gail Eisnitz's, *Slaughterhouse: The Shocking Story of Greed, Neglect, and Inhumane Treatment inside the U.S. Meat Industry* ... from which it in fact comes! Yes, the fragment does *not* come from Stephen S. Brandom's 1999 novel *Darkness Over Chicago*. Truth be told, no such novel even exists.

We hope the readers will forgive us playing games with them, but our deception was meant to illustrate the point we were making. That is, some texts may be equally well perceived as both fictional and non-fictional, and how exactly they will be perceived can be influenced by priming the reader in a particular way. The above trick was also meant to illustrate something more. It was meant to allow the readers of this book to have an inside perspective on the experiment we are about to describe, as the extract comes from the very text we employed. And the text *really* is (the reader can rest assured we are being sincere this time) a story included in Eisnitz's *Slaughterhouse*.

From The Slaughterhouse *to* The Jungle *(and back)*

One reason we chose that particular narrative was of course the fact that, while a journalistic report, it could definitely be read as a detective story. But we chose it also because the book it comes from has been reported to have a considerable impact on attitudes toward farm animals in the USA and around the world, and because it has been recommended by vegetarian activists as a "great resource" for animal narratives that can "hook the readers' attention" (Breier and Mangels 2001, 46). Indeed, it is hard to disagree with the philosopher Peter Singer that, although the facts Eisnitz writes about in her book are gruesome, she nevertheless writes about them "superbly" (Eisnitz 2007 back cover).

The readers of our book already know Eisnitz as the source behind the *Washington Post* story "They Die Piece by Piece" that caused such uproar in the USA that it was heard in the American Senate. Having read that article, Senator Robert Byrd

> was moved to make an impassioned speech on the Senate floor: 'Animal cruelty abounds. It is infuriating. The barbaric treatment of helpless, defenseless creatures must not be tolerated ... Such insensitivity is insidious and dangerous. Life must be dealt with humanely in a civilized society'.
>
> (Eisnitz 2007, 301)

Perhaps more importantly, Robert Byrd put his money where his mouth was, expending considerable efforts to convince the Senate to allocate millions of dollars for the improvement of farm animal welfare (301–2).

But *Slaughterhouse* came before that. Originally published in 1997, the book was the result of a systematic investigation into the workings of the American meat industry which Eisnitz conducted over several years. The investigation included undercover penetration of slaughtering facilities, studying piles of documents, as well as interviewing numerous workers with thousands of hours of experience on the killing floor. The events and data it recounted were real, which made the picture painted in the book all the more shocking. What was going on behind the walls of farms, abattoirs, and meat facilities was so gut-wrenching that it would otherwise have been hard to believe had it not been true.

In this way, and as Eisnitz herself suggests (2007, 21), *Slaughterhouse* could be compared to one of the most famous American books of the first half of the twentieth century, Upton Sinclair's 1906 *The Jungle* (2003; Halley 2012, chap. 7). That book similarly exposed the American public to the truth about the conditions in which their meat was produced, and that truth was also horrifying, outrageous, and surprising. First, it turned out that the meat Americans bought on a daily basis was tainted with the plight of thousands of people who helped to produce it, the ubiquitous low-wage, poor meatpackers, often recruited from both legal and illegal immigrants. Second, that meat was tainted in a more literal way too. It would often come from animals infected with all sorts of diseases, and even in those cases where it came from animals who were healthy, those animals were killed in such a way, and in such conditions, that the end product had to be unhealthy itself. All these circumstances, Sinclair argued, were the function of the model of capitalist production in the USA at the time, where profit trumped all other considerations (Bloom 2010, 3–47). In this particular case, it trumped consideration for the health of meat packers, and meat consumers. Caring about such things costs money, after all.

Unfortunately, 90 years after *The Jungle*, Eisnitz could accuse the meat industry of similar faults and explain them in a similar way. She documents throughout the book that many of the meatpackers are immigrants, some of them illegal, and that the conditions in which they work are extremely bad. According to the official statistics obtained by the author,

> with nearly thirty-six injuries or illnesses for every one hundred workers, meat packing is the most dangerous industry in the United States. In fact, a worker's chances of suffering an injury or an illness in a meat plant are six times greater than if that same person worked in a coal mine.
>
> (Eisnitz 2007, 271)

And her in-depth investigation provided Eisnitz with gruesome details that make those data not only more concrete but also even more shocking.

Moreover, despite all the legal instruments that had been introduced after the publication of Sinclair's book in order to protect consumers from meat of insufficient quality, Eisnitz found that the products bought by today's consumers are a potential health hazard too, one of the reasons being that they are often contaminated with dangerous bacteria such as Campylobacter, "the number-one cause of gastroenteritis in the United States, causing hundreds of deaths each year":

> According to the National Academy of Sciences, studies of market-ready chickens found Campylobacter on up to 82 percent [of store-bought chickens]. And in a survey of fifty brand-name broilers in Georgia, a government researcher found 90 percent contaminated with [it].
>
> (Eisnitz 2007, 177)

Again, this seemed to be just the tip of an iceberg, and as it was the case with the realities described in *The Jungle*, according to Eisnitz, the major source of the problem she described was again "greed," a term which aptly figures in the subtitle of her book.

But there are also significant differences between the two books. For one, Sinclair's text is a novel. It was advertized as such and perceived accordingly by his readers, whereas Eisnitz's book is a journalistic report and this is how it has been presented on the book market. For two, if in writing *The Jungle* Sinclair "aimed at the public's heart" (Bloom 2010, 6), he meant to open it mainly (if not exclusively), to the plight of the *humans* harmed by the workings of the meat industry. Eisnitz on her part focused instead predominantly on the suffering of non-human animals.

She focused on the millions of animals who are slaughtered in inhumane ways every day. She wrote about cows being "hoisted upside down and butchered-while still alive." She wrote about pigs "routinely dragged into narrow alleyways between pens where they were provided no food or water and were left to die slowly of disease, starvation, and dehydration." She wrote about horses beaten with "pipes," kicked, and stuck "with knives" (2007, 138, 296, 306). It is one of her stories about the treatment of horses that we used in our experiment. It was extracted from Chapter 7 of *Slaughterhouse* and centers around an interview Eisnitz conducted with somebody we already know – Mr. Parrish, a state penitentiary inmate who had been involved in the illegal slaughtering of horses.

A Tale of Two Experiments, or How to Learn from One's Mistakes

The participants in our experiment were 114 high school students aged between 18 and 19 (58 women) divided into three groups. There was a control group, who read a narrative placebo, i.e., a story with a topic neutral from the point of view of our study. And then there were two groups who read the horse story extracted from Eisnitz's book. One of these latter groups was induced to think that the story was fictional, while the other that it was not, but otherwise the experimental conditions for them were the same. We manufactured this difference using what is sometimes called the frame of a text, or "paratext" (Genette 1997), where these terms are understood as denoting any textual material that accompanies the main text and gives the reader information about it. Some common examples of the textual frame are titles, tables of contents, blurbs, and the like. In our case, the frame was a brief introductory passage that preceded the story and differed between the two experimental groups. The information we gave to the group who was supposed to see the story as fictional was the same as the fake one we gave to you, the readers, above, and read:

> The following fragment comes from Stephen S. Brandom's, detective novel *Darkness Over Chicago* (1999). The main protagonist of the novel, detective John McDowell, is conducting an investigation about the ties which the New York and Chicago Mafia have with various business circles, including art market, gastronomy and food industry. In the following fragment, the detective arrives at a state prison to meet with an inmate who was involved in illegal meat trade.

The people in the "non-fiction" experimental condition, in turn, were told the truth: that the story they were about to read comes from a "journalistic" book by Gail Eisnitz about the workings of the American meat industry, that it portrays "real events," is based on interviews with meat industry workers, and that it was widely discussed in the newspapers following its publication.

The main text was the same for both groups. It began with the sentences we quoted above ("Steve Parrish finally entered and was directed to where I was seated..."), and then it included various details about the illegal activities which Mr. Parrish had been involved in, including selling non-inspected meat of stolen horses to restaurants as beef, and the like. It also included nauseating details of how the slaughterers killed the horses, as brutal and sickening as something you could find in a really noir detective story:

> The buzzer rang. After the prisoner count, I asked Parrish how the slaughter went for horses who walked into the knocking box.
>
> "There's a certain way to shoot or knock an animal," he said. "I seen them shoot them five times, hit them all in the eye. Hit them in the neck. I seen horses get shot wrong and get right back up and walk around the kill floor, kind of dazed. And they run up on them and just hit them with the knife in the neck, anywhere, and just let them suffer, walk around bleeding."
>
> "Sometimes they can't get close enough with the knocking gun," he continued. "It didn't work right sometimes, sometimes the gun gets wet, gets blood up in it, and it don't shoot. The boss tells us, 'run and cut his throat.' I've seen my boss grab a knife and run and cut its throat."
>
> "What about the inspector?" I asked. "Does he ever see any of this?"
>
> "Yes." (Eisnitz 2007, 139)

And so the text went on, consisting of over 1,000 words altogether.[1] After they had finished reading it, the subjects were asked to complete our questionnaire, and, once they had done that, they were thanked for their participation.

By comparing the results of those two groups, we were then able to measure whether the perceived fictionality of the text had an impact on the way it influenced attitudes toward animal welfare. But note that in order to establish whether our narrative had a statistically significant influence on those attitudes in the first place we had to compare the results of our two experimental groups with the results of the control group, who read a narrative placebo. In this experiment, the placebo was a journalistic story about the then-recent discovery of the Higgs boson particle that had been published in a fairly popular online Polish daily.[2]

With such an experimental design everything should have worked smoothly. But it did not, and here is one of the failures that we earlier promised to admit to. The failure had a lot to do with the fact that we attempted to perform our experiment at a high school renowned for its academic achievements as well as for its sharp, curious, and proud students. Unfortunately, they proved too sharp, curious, and proud for the design we prepared. Having received the text, some of those in the fictional-frame condition group immediately began to google the author and the title of the novel, and were quick to realize that no such author and novel existed. They then proudly announced this during the running of the experiment, and explicitly chided our confederates for underestimating their intelligence. Of course, the experiment was a failure. Everything depended on our subjects believing the paratextual frame to be true, and it was compromised as fake. Admittedly, some of the students might have still believed that while not part of the novel, the fragment was nevertheless fictional, but we could not be sure of that. All our efforts were for no reward.

We had to rerun the study with a better paratextual frame and at a different school. We could be sure that many people at *that* school quickly learned about how some silly academics had been made fools of, and that if we returned there, our subjects would most likely treat us with caution. From that time on we would never again underestimate the power which smart phones could have in the hands of our participants. Our revised paratextual frame was supposed to neutralize that power. In the fictional condition, it presented the horse story as taken from an actual novel whose title, author, and publisher could be easily found on the web. As could the summaries and reviews of that novel, all of which, to our best knowledge, were consistent with the fragment we asked the subjects to read.

Our new paratextual description read as follows:

> The following fragment comes from Ed McBain's detective novel *Another Part of the City* (1999). The main protagonist of the novel, detective Reardon, is conducting an investigation about the ties which the New York and Chicago Mafia have with various business circles, including art market, gastronomy and food industry. In the following fragment, the detective arrives at a state prison to meet with an inmate who was involved in illegal meat trade.

We tested our improved design at another high school on 114 students, including 58 women, and this time no suspicions were expressed – neither spontaneously nor when the confederates asked the subjects what they thought about the study after it had been over. We could then, prima facie, trust our data from this re-run of the experiment.

Sometimes Truth Is Stranger than Fiction, or Our Results

After we submitted the data to statistical analysis, they showed, first of all, that, similarly to the Krajewski study, women expressed generally more pro-animal welfare attitudes than men ($p < 0.001$, $\eta^2 = 0.12$) and that participants who stated that they had a pet at home expressed more pro-animal welfare attitudes ($p < 0.02$, $\eta^2 = 0.05$). As in the Krajewski study, animal stories were also shown to improve attitudes toward animals. In comparison with the control group, the impact of both versions of Eisnitz's story taken together was statistically significant, $p < 0.02$, and the value of η^2 was even higher than in the Krajewski study, equaling 0.07. This meant that the size of the effect was slightly bigger, or that that particular story had an even stronger impact on attitudes toward animal welfare. This was an exciting and promising result. We seemed to be off to a good start with our laboratory experiments.

But what about the supposed difference between the experimental groups? Before we present our data, we would like to ask our readers what *they* think. Which kind of story is more likely to change attitudes toward animals, one that you believe to depict actual or fictional animals? We suppose that most of our readers would say that the former. After all, isn't real suffering more important than fictional harm? And doesn't it make more sense to change one's mind about real social issues on the basis of factual information rather than fabricated stories? This view is certainly intuitive and, moreover, supported by psychological data. For instance, in one of his experiments, C. Daniel Batson (2002) showed that when his participants believed that the drug-addicted protagonist of a story they read was fictional, their attitudes toward drug addicts improved to a significantly lesser degree than was the case with those participants who were induced to believe that the story was real.

On the other hand, there are also theories which suggest that the opposite might be the case, i.e., that a narrative perceived to be fictional might have a stronger impact than one perceived to be factual. This alleged effect is attributed to the fact that fictional discourse provides us with a space in which we can give in to our feelings relatively safely, that is without having to consider the practical consequences of doing so ("here is social life without obligation, meeting without responsibility" (Oatley 1999, 445)), whereas in real life, one is always reminded that a bleeding heart can cost a bloody lot. In other words, we might hypothesize that the subjects in the fiction condition could afford to yield to their emotions to a greater degree than those in the non-fiction condition (Oatley 2002, 63–64; cf. Shusterman 2001) and that this could have made them more susceptible to attitudinal change.[3]

With this in mind, let us return to our data. Which of the above hypotheses does it support? Which of our experimental groups showed better attitudes toward animal welfare? The one whose members thought the story was fictional or the one whose members thought it was a piece of journalism? The answer is, neither. That is, while the story significantly improved the readers' attitudes as compared to the control group, there was no significant difference between the degree of that influence between the two groups. Sounds strange? We bet it does. It definitely feels that there *should* be a difference in impact between fiction and truth, shouldn't there?

Well, sometimes truth is stranger than fiction. In fact, our results are not that extraordinary, being comparable with several other experiments showing that psychological impact of narratives does not always depend on their real-world status, that is, on whether they are perceived as fictional or not. For instance, that status has turned out not to have any significance for the influence of stories on empathic understanding toward people undergoing severe grief or depression and on prosocial behavior on their behalf (Koopman 2015). Perceived fictionality has been also reported, in a widely cited study by Melanie C. Green and Timothy C. Brock (2000), to lack any impact on how stories affect the readers' beliefs. As Green and Brock summarize their findings, it appears that

> once a reader is rolling along with a compelling narrative, the source [which the narrative draws on] has diminishing influence. In this fashion, the belief positions implied by the story might be adopted regardless of whether they corresponded with reality.
>
> (2000, 719)

Unfortunately, Green and Brock do not try to provide any explanation for this phenomenon, and since we believe that it is a scholar's duty to provide one whenever he or she reports on a truth that is stranger than fiction, here is our hypothesis. First of all, note that both in the case of the studies referenced above and our experiment, the experimental stories concerned matters that undoubtedly elicit strong emotions: depression, grief, the inhumane slaughter of horses, and even, in the Green and Brock study, being "brutally stabbed by a psychiatric patient." Second, note that according to current neurology matters of this kind are processed by a different part of the brain than the information about whether the story that touches on them is true. That is, the emotional aspects of a narrative event are processed by the limbic system, while it is the function of the cerebral cortex to adjudicate when, where, and if the event actually took place. The limbic system is much older and more basic than the cerebral cortex in evolutionary terms, which allows it to often override what the cortex suggests. To use an apt metaphor proposed by the psychologist Jonathan Haidt (cf. Jonathan Haidt 2006; Haidt 2001), the relation between the two can be likened to the one between an elephant and its human rider:

> Perched atop the Elephant, the Rider holds the reins and seems to be the leader But the Rider's control is precarious because the Rider is so small relative to the Elephant. Anytime the six-ton Elephant and the Rider disagree about which direction to go, the Rider is going to lose. He's completely overmatched.

Most of us are all too familiar with situations in which our Elephant overpowers our Rider. You've experienced this if you've ever slept in, overeaten, dialed up your ex at midnight, procrastinated, tried to quit smoking and failed, skipped the gym, gotten angry and said something you regretted, abandoned your Spanish or piano lessons, refused to speak up in a meeting because you were scared, and so on.

(Heath and Heath 2010, 7)

Apparently, the "so on" includes also all the times when you got really scared reading a horror story or spent a lot of time worrying what will happen to your favorite character in the next episode of your favorite TV series instead of focusing on practical matters that were of much more urgent importance. It would seem, then, that what happens at such times is this: however much your cerebral cortex insists that your reaction to the story in question is inappropriate and unreasonable, since the force behind the reaction is something akin to a six-ton monster, such admonitions must be in vain.

This may be seen as a disadvantage of our brain, but according to some evolutionary explanations of the origin of storytelling, our willingness to treat stories seriously no matter what their source is in fact an evolutionary adaptation. For, as argued by evolutionary scholars, fictional stories have the same role in our lives as play in general has in the life of animals. It is a simulation which allows us to practice certain skills that are useful in reality. "Ordinary play allows animals to extend and refine their competence in standard species behaviors to the point where their skills offer a new freedom that may be crucial in situations like attack, defense, or rearing offspring," writes Brian Boyd. "The special cognitive play of art," he proposes "allows humans to extend and refine key cognitive competences" (2009, 190).

Steven Pinker concurs:

Intelligent systems often best reason by experiment, real or simulated: they set up a situation whose outcome they cannot predict beforehand, let it unfold according to fixed causal laws, observe the results, and file away a generalization about how what becomes of such entities in such situations. Fiction, then, would be a kind of thought experiment, in which agents are allowed to play out plausible interactions in a more-or-less lawful virtual world.

(2007, 172)

Arguably, from an evolutionary point of view, the most important of such thought experiments would consider things that are the most important for the survival of our species, which are again precisely those matters which elicit in us the strongest emotional responses, or make our elephants run: death, disease, pain, power, and sex.

Whether our hypothesis is true or not, the particular feature of narratives it tries to explain is very important from the point of view of our project. Recall that one reason why we hypothesized literary narratives may be a valuable tool in influencing the public's concern about animal welfare is that thanks to their aesthetic appeal they can make the topic of animal plight interesting even to those people who would otherwise refuse to know more about it out of the fear that this could cause them too much distress or make them feel guilty. We think it is reasonable to assume that that capacity of literary narratives would be even more pronounced in the case of literary stories that are perceived as fictional. If you think that a story of animal suffering is fictional, you should definitely experience such fears to a lesser degree than you would in the case of a story you thought was real.[4] And there are countless stories of this kind aside from those that we mentioned, from Disney's *Bambi*, which allegedly had a huge impact on attitudes toward hunting in the USA, to more serious works such as Steinbeck's *The Red Pony* (Steinbeck 1994; Molloy 2011, 30; Rudy 2011, 201). The good news brought by our results is that such a hypothetical effect would not come at the price of decreased persuasiveness. As we have just shown, the attitudinal impact of an animal story does not depend on whether you perceive it to be fictional or not.

Notes

1 The exact number of words was 1,167. Note that in the experiment we used a Polish translation of the text, which we prepared ourselves.
2 The text met our criteria for a control narrative to a very satisfying degree. Its topic was neutral from the point of view of our study, it possessed literary qualities, and was additionally suitable for use in the whole series of our planned laboratory experiments, in which we were going to deploy experimental narratives of diverse types, including fiction, autobiographical essay, journalism, and the like (Bożek 2012). Note that the placebo used in this study is not the same as the one deployed in the experiment described in Chapter 2, where we used a fragment of *The Lord of the Numbers*. The reason for this is purely contingent: it so happened that we had to begin our laboratory studies before Krajewski provided us with the manuscript of his novel.
3 Although perhaps less intuitive, this hypothesis, too, might be supported by experimental data, even if indirectly so. In his 2013 study, the American psychologist Dan R. Johnson studied attitudinal change in "individuals low in dispositional perspective-taking," i.e., people whose capacity to "put themselves in someone else's shoes" is below average. He found that such people additionally tend to express higher levels of what psychologists call intergroup anxiety. They are anxious to interact with members of outgroups – other races, classes, ethnic and sexual minorities. But his results also suggest that fiction can offer such people "a safe haven" from intergroup anxieties. Johnson inferred this from the fact that in his experiment, in which he studied the impact of fictional narratives on attitudes toward Arabs and Muslims, "implicit prejudice was reduced the most for individuals low in dispositional perspective-taking" (2013, 588).

4 Consider, for instance, that some people who easily watch action movies where heads are severed, arms broken, and blood spilt have problems watching a documentary footage narrating the same kind of bodily harm, however aesthetically attractive the footage might be. One reason for this is most probably their fear of distress and this is precisely one of the reasons why such footage is rarely shown on TV, in contrast to bloody action movies, which seem to be one of TV's main attractions.

Works Cited

Argentino, Concetta M., Aline H. Kidd, and Karen Bogart. 1977. "The Effects of Experimenter's Sex and Subject's Sex on the Attitudes toward Women of Fraternity, Sorority, and Mixed-Dormitory Residents." *Journal of Community Psychology* 5 (2): 186–88. doi:10.1002/1520–6629(197704)5:2<186::AID-JCOP2290050215>3.0.CO;2-A.

Batson, C. Daniel, Johee Chang, Ryan Orr, and Jennifer Rowland. 2002. "Empathy, Attitudes, and Action: Can Feeling for a Member of a Stigmatized Group Motivate One to Help the Group?" *Personality and Social Psychology Bulletin* 28 (12): 1656–66. doi:10.1177/014616702237647.

Bloom, Harold, ed. 2010. *Upton Sinclair's The Jungle.* Bloom's Modern Critical Interpretations. New York: Bloom's Literary Criticism.

Boyd, Brian. 2009. *On the Origin of Stories: Evolution, Cognition, and Fiction.* Cambridge, MA: Belknap Press of Harvard University Press.

Bożek, Jakub. 2012. "Ta cholerna cząstka." *Krytyka Polityczna*, July 8, 2012. http://krytykapolityczna.pl/nauka/ta-cholerna-czastka/.

Breier, Davida Gypsy, and Reed Mangels. 2001. *Vegan & Vegetarian FAQ: Answers to Your Frequently Asked Questions.* Baltimore, MD: Vegetarian Resource Group.

Eisnitz, Gail A. 2007. *Slaughterhouse: The Shocking Story of Greed, Neglect, and Inhumane Treatment inside the U.S. Meat Industry.* Amherst, NY: Prometheus Books.

Genette, Gérard. 1997. *Paratexts: Thresholds of Interpretation.* Literature, Culture, Theory 20. Cambridge; New York: Cambridge University Press.

Green, Melanie C., and Timothy C. Brock. 2000. "The Role of Transportation in the Persuasiveness of Public Narratives." *Journal of Personality and Social Psychology* 79 (5): 701–21.

Haidt, Jonathan. 2006. *The Happiness Hypothesis: Finding Modern Truth in Ancient Wisdom.* New York: Basic Books.

———. 2001. "The Emotional Dog and Its Rational Tail: A Social Intuitionist Approach to Moral Judgment." *Psychological Review* 108 (4): 814–34.

Halley, Jean O'Malley. 2012. *The Parallel Lives of Women and Cows: Meat Markets.* Critical Studies in Gender, Sexuality, and Culture. New York: Palgrave Macmillan.

Harris, Sandra. 1971. "Influence of Subject and Experimenter Sex in Psychological Research." *Journal of Consulting and Clinical Psychology* 37 (2): 291–94. doi:10.1037/h0031780.

Heath, Chip, and Dan Heath. 2010. *Switch: How to Change Things When Change Is Hard.* 1st ed. New York: Broadway Books.

Hendrick, Clyde, and Russell A. Jones. 2013. *The Nature of Theory and Research in Social Psychology.* Burlington, NJ: Elsevier Science.

Johnson, Dan R., Daniel M. Jasper, Sallie Griffin, and Brandie L. Huffman. 2013. "Reading Narrative Fiction Reduces Arab-Muslim Prejudice and Offers a Safe Haven From Intergroup Anxiety." *Social Cognition* 31 (5): 578–98. doi:10.1521/soco.2013.31.5.578.

Koopman, Eva Maria (Emy). 2015. "Empathic Reactions after Reading: The Role of Genre, Personal Factors and Affective Responses." *Poetics* 50 (June): 62–79. doi:10.1016/j.poetic.2015.02.008.

Mar, Raymond A., Keith Oatley, Jacob Hirsh, Jennifer dela Paz, and Jordan B. Peterson. 2006. "Bookworms versus Nerds: Exposure to Fiction versus Non-Fiction, Divergent Associations with Social Ability, and the Simulation of Fictional Social Worlds." *Journal of Research in Personality* 40 (5): 694–712. doi:10.1016/j.jrp.2005.08.002.

Molloy, Claire. 2011. *Popular Media and Animals*. Palgrave Macmillan Animal Ethics Series. Houndmills, Basingstoke; New York: Palgrave Macmillan.

Nichols, Austin Lee, and Jon K. Maner. 2008. "The Good-Subject Effect: Investigating Participant Demand Characteristics." *The Journal of General Psychology* 135 (2): 151–66. doi:10.3200/GENP.135.2.151-166.

Oatley, Keith. 1999. "Meetings of Minds: Dialogue, Sympathy, and Identification, in Reading Fiction." *Poetics* 26 (5–6): 439–54.

———. 2002. "Emotions and the Story Worlds of Fiction." In *Narrative Impact: Social and Cognitive Foundations*, edited by M. C. Green, J. J. Strange, and T. C. Brock, 39–69. Mahwah, NJ: Lawrence Erlbaum Associates Publishers.

Pinker, Steven. 2007. "Toward a Consilient Study of Literature." *Philosophy and Literature* 31 (1): 162–78. doi:10.1353/phl.2007.0016.

Rudy, Kathy. 2011. *Loving Animals: Toward a New Animal Advocacy*. Minneapolis: University of Minnesota Press.

Shaughnessy, John J., Eugene B. Zechmeister, and Jeanne S. Zechmeister. 2012. *Research Methods in Psychology*. New York: McGraw-Hill.

Shusterman, Richard. 2001. "Art as Dramatization." *The Journal of Aesthetics and Art Criticism* 59 (4): 363–72.

Sinclair, Upton. 2003. *The Jungle: An Authoritative Text, Contexts and Backgrounds, Criticism*, edited by Clare Virginia Eby. 1st ed. A Norton Critical Edition. New York: Norton.

Steinbeck, John. 1994. *The Red Pony*. Penguin Twentieth-Century Classics. New York: Penguin Books.

Strange, Jeffrey J. 2002. "How Fictional Tales Wag Real-World Beliefs." In *Narrative Impact: Social and Cognitive Foundations*, edited by Melanie C. Green, Jeffrey J. Strange, and Timothy C. Brock, 262–86. Mahwah, NJ: L. Erlbaum Associates.

Webster, Murray, and Jane Sell, eds. 2007. *Laboratory Experiments in the Social Sciences*. Amsterdam; Boston, MA: Academic Press/Elsevier.

Williams, Sue Winkle, Shirley M. Ogletree, William Woodburn, and Paul Raffeld. 1993. "Gender Roles, Computer Attitudes, and Dyadic Computer Interaction Performance in College Students." *Sex Roles* 29 (7–8): 515–25. doi:10.1007/BF00289325.

4 Does It Matter How It Is Told?

On Species, Stylistics, and Voices

I Am the One Who Knocks

If one were to name the most successful crime story of recent years, regardless of the medium, then judging by the size of its audience, the awards it has received, and the cultural impact it has made, one would probably nominate the American TV Series *Breaking Bad*, created by Vince Gilligan. Given its rather unusual main premise, the success of the series is admittedly quite remarkable. It tells of an amiable chemistry teacher named Walter White who has a regular, peaceful, and unexceptional life which he shares with his loving family – a wife and son. Or rather he had such a life, as the story begins when he is diagnosed with incurable lung cancer with only a few years left to live. Walter loves his family very much, so for him the worst part of the situation is not so much the fact that he will not survive his illness, but how his family will survive once he has gone. His wife, Skyler, is an intelligent woman who unfortunately does not have a real job, only some rather poor business ideas that fail to provide a significant living. His son, though a clever young man, would not be easily able to help her as he is still a high school student, not to mention that he has cerebral palsy. And then there is the second child, still in Skyler's womb, to be born in a few months.

Obviously, the financial prospects for the family are grim. With his salary of a high school teacher, Walter could not gather much in the way of savings and so cannot secure their future. What is he to do to change that? Rob a bank? Not exactly. Instead, a series of events leads him down a different criminal path. He utilizes his chemical expertise (far exceeding that of an average chemistry teacher) to produce the best methamphetamine on the market in the hope that by selling a sufficient amount of it he will be able to secure a financially stable life for his loved ones. He achieves that, and far more besides. At first almost unwittingly, and then with increasing ruthlessness, he becomes one of America's most powerful drug lords and somehow manages to keep that secret from his family for an impressively long time.

That story is of course morally dubious and rather improbable, but nevertheless, or perhaps partly because of that, it appealed to millions of

DOI: 10.4324/9780429061424-5

viewers, extending into as many as five seasons showered with all kinds of awards and media attention. It has also become the object of a cult following with fans celebrating and scrutinizing the twists and turns of its plot, the psychology of its characters, and even particular scenes. One of the most famous of the latter is the so-called "I am the one who knocks" monologue. It has been discussed in all sorts of media and all over the Internet (Snierson 2016). Of the dozens of videos on YouTube which contain that scene, one has five million views while another over a million, not to mention that there are also numerous videos with alternative renditions of the monologue, including by Samuel L. Jackson, with two million views (Couch 2013a).

The scene takes place in episode 6 of the fourth season, months after Skyler had learned about the criminal side of Walter's life (though at that point she is still in the dark about many of its aspects) and shortly after an event that made her believe that Walter is in danger of being murdered. The two are having a conversation and she urges him to turn himself in to the police by saying:

> I know what it could do to this family. But if it's the only real choice we have – if it's either that, or you getting shot when you open your front door …You are not some hardened criminal, Walt! You are in over your head. That's what we tell them. That's the truth!
> … A schoolteacher with cancer, desperate for money, unable to even quit … You told me that yourself, Walt. Jesus, what was I thinking? Walt, please! Let's both stop trying to justify this thing, and admit you're in danger.

While she talks, Walter is shaking his head, apparently deliberating in his mind whether to keep the image she harbors of him intact and let her worry, or shatter that image and calm her worries down. He decides on the latter, asking her:

> Who are you talking to right now? Who is it you think you see? Do you know how much I make in a year? I mean, even if I told you, you wouldn't believe it. You know what would happen if I suddenly decided to stop going in to work? A business big enough that it could be listed on the NASDAQ goes belly up. Disappears. It ceases to exist without me. No – you clearly don't know who you're talking to. So, let me clue you in. I am not in danger, Skyler. *I am the danger.* A guy opens this door and gets shot, and you think that is me. No. *I am the one who knocks!*

> (Gilligan et al. 2009)

Then he leaves the room, leaving his wife dumbstruck and the viewers in awe at this rhetorical gem, which has become an immediate classic – discussed, analyzed, and praised (Couch 2013b; Mittell 2015, 161–62). That status has been noticed and even become the subject of gentle satire. Published on the website McSweeney's (Spadanuta 2013), the satire half-jokingly puts the monologue in the context of the literary canon, presenting its various versions as hypothetically written by famous authors.

And so Jane Austen would write it thus:

> "I'm the person who gentle folk hear after dinner, what strikes fear in their drawing rooms," our heroine overheard the balding gentleman in the dark hat and spectacles remark to his astonished wife. "Perhaps we should take to Bath this summer," the wife replied, changing the subject.
>
> (Spadanuta 2013)

And how about a hypothetical version by John Steinbeck?

> Toast crumbs mingled with butter and the Albuquerque sand in his beard. The auburn hairs engulfed the particles in a flame that would never breathe or grow. He had taken his glasses off but they left marks on his temples, like the skid marks of a teenage drag race in the Dog House parking lot. "I'll be the one who's comin' round to 'em," he said, his spittle dripping into the carpet fibers.
>
> (Spadanuta 2013)

Or Toni Morrison?

> Into the fading lights of his wife's shuttering eyes he stared. "Knocking. Answering. Death."
>
> (Spadanuta 2013)

And finally revel in this little gem in Hemingway's style:

> "I knock," Walt said. That was all.
>
> (Spadanuta 2013)

Of course, the above is not a serious literary exercise as it exaggerates and otherwise deforms the discursive qualities of these authors' prose. But precisely because these authors are so ripe for satirization, it highlights how important an author's stylistic choices are in determining the impact their stories can have on their audience (Leech and Short 2007). What if Hemingway's tales were told in a more lyrical way? And Steinbeck's looked more like telegraphic newspaper reports? What if Jane Austen did not narrate her stories from the point of view of the main protagonist ("our heroine"), giving her narrators more room to breathe?

Would their novels and stories work the way they do? And if so, what would that tell us about narratives in general? Perhaps there are certain regular psychological patterns at work here? Maybe there are certain stylistic devices that are more conducive to bringing about certain kinds of effects?

This brings us to our animal narratives. Note that two of the most impactful animal stories of the nineteenth century, *Beautiful Joe* and *Black Beauty,* were animal stories not only in the sense that they concerned animals. They were also stories told from the point of view of an animal and told *by* it. Would they have influenced the public the way they did had they been written entirely from a third-person perspective or did the key to their success lie partly in the situated animal perspective from which they were told? Contemporary animal advocates and sympathizers probably thought the latter as many other narratives they used involved that narrative strategy as well (Cosslett 2006; Boggs 2013; Elick 2015).

In this chapter, we will investigate if they were right, and we will also tackle some other questions related to how animal stories are told, or to their discursive characteristics. One such question is whether the impact of a story depends on its being accompanied by arguments which explicitly articulate its message. Another is whether that impact might be strengthened if the story contained indirect normative hints as to how the reader should judge, or feel about, the events described. These questions are important for this book because such strategies have often been used in animal stories, but they are important even beyond that specific focus in that there have been numerous voices in literature and literary criticism arguing that the moral power of literature, including its superiority over sermons and propaganda, lies precisely in that it avoids making its points explicit or even suggesting how the reader should think (Foulkes 1983; Oatley 2011, 174). Some would even say that this power lies in its capacity to simply tell it like it is, in the minutest details that no other discourse can convey. The idea here is that the moral power of literature is precisely the function of its being something that can be so easily shattered and ignored, a mirror that reflects reality (Dickstein 2005; Mazzoni 2017, 187–89). Whether that idea itself reflects reality is still not clear. We hope that our experiments will contribute to changing that.

Alice Walker's "Am I Blue?": Or Stories and Arguments

Is It OK for a Story to Tell You What Its Point Is? From Hesiod to Ta-Nehisi Coates

Although there are many who think that literary stories should speak for themselves, that it is bad taste to include within them arguments articulating their point, historically this has not always been a popular view. Indeed, there was a time when those literary stories which had a moral point were indeed *expected* to lay it out clearly, and some of those which did definitely belong to the canon of Western literature. Consider, for instance, Hesiod's *Works and Days* or Virgil's *Georgics* or *The Pilgrim's Progress* by John Bunyan.

But even today authors who want their literary stories to effect a moral change sometimes articulate arguments within them, and the results are not necessarily anachronistic. Indeed, some such stories are as fresh and engaging as could be. One good example would be the 2015 winner of the National Book Award, *Between the World and Me* by Ta-Nehisi Coates (2015). Written as a series of letters to the author's son that condemn the racial injustice of American society, the book is partly a story of the author's growing up black in Baltimore, and partly a history of race relations in America. But it is more than that as it expands on those two intertwined narratives with a host of arguments of different sorts – sociological, ethical, psychological, and more. Consider this small episode:

> When I was six, Ma and Dad took me to a local park. I slipped from their gaze and found a playground. Your grandparents spent anxious minutes looking for me. When they found me, Dad did what every parent I knew would have done—he reached for his belt. I remember watching him in a kind of daze, awed at the distance between punishment and offense. Later, I would hear it in Dad's voice—"Either I can beat him, or the police." Maybe that saved me. Maybe it didn't.
>
> (Coates 2015, 16)

And now consider how this individual story is situated by the author in the context of similar stories of other African-American children, and how the social significance of these narratives is then explained with the help of arguments:

> All I know is, the violence rose from the fear like smoke from a fire, and I cannot say whether that violence, even administered in fear and love, sounded the alarm or choked us at the exit. What I know is that fathers who slammed their teenage boys for sass would then release them to streets where their boys employed, and were subject to, the same justice. And I knew mothers who belted their girls, but the belt could not save these girls from drug dealers twice their age.
> ...
> To be black in the Baltimore of my youth was to be naked before the elements of the world, before all the guns, fists, knives, crack, rape, and disease. The nakedness is not an error, nor pathology. The

nakedness is the correct and intended result of policy, the predict-
able upshot of people forced for centuries to live under fear. The law
did not protect us. And now, in your time, the law has become an
excuse for stopping and frisking you, which is to say, for furthering
the assault on your body. But a society that protects some people
through a safety net of schools, government-backed home loans, and
ancestral wealth but can only protect you with the club of criminal
justice has either failed at enforcing its good intentions or has succeeded at something much darker. However you call it, the result
was our infirmity before the criminal forces of the world. It does not
matter if the agent of those forces is white or black— what matters
is our condition, what matters is the system that makes your body
breakable.

(17–18)

There are many more passages of this kind in the book, and there are
many more works like this one, some of which concern the plight of
animals. In the experiment described in this subchapter, we wanted
to see whether the discursive strategy they employ – that is, mixing
narrative with arguments (or with what is sometimes called expository
discourse) – makes the impact of narratives stronger.

It so happens that one of the most famous of contemporary literary
texts that attempt to make an impact on attitudes toward animals is
precisely of this narrative-argumentative kind. It also happens that, just
like *Between the World and Me*, it was written by an African-American
author and concerned the status of that racial minority. Its title is "Am I
Blue?" and it was penned by Alice Walker (Walker 2010). It is this text
that we used in our experiment on the role of arguments in the attitudinal impact of narratives. But before we explain what we did with the
text and did to it, a word is due on the text itself.

From The Color Purple *to* "Am I Blue?"

Alice Walker has been already mentioned in this book as the author of
the 1982 epistolary novel *The Color Purple*. The novel earned her the
Pulitzer Prize in 1983, was adapted for screen by Steven Spielberg in
1985, and is sometimes assigned at school, in part, for the purpose of
fighting prejudice against people of African descent, as well as for its literary quality (Ngo and Kumashiro 2014). This is the work that Walker
is most widely known for, while racial politics is *the* topic she is most
frequently associated with. But she also happens to be an animal liberation advocate and has expressed that in her writings as well, including in
"Am I Blue?" (Hooker 2005).

The main theme of "Am I Blue?" is to draw a parallel between the
plight of nonhuman animals and the plight of African Americans and

other human minorities. While that parallel has often provoked controversy (it has been even called "the dreaded comparison"; Spiegel 1996), it is also intertwined in the theoretical background of the animal liberation movement. In fact, one of the key notions in that theory – speciesism – was coined precisely in order to stress some fundamental similarities between the ideology of racism and the ideology with which our society justifies its oppression of animals (Singer 2009; Gruen 2011, 54).

The similarities Walker focuses on in her essay concern all sorts of ideological assumptions about the capacity of people from minority groups and nonhuman animals to suffer. "There are those," she argues,

> who never once have even considered animals' rights: those who have been taught that animals actually want to be used and abused by us, as small children "love" to be frightened, or women "love" to be mutilated and raped … They are the great-grandchildren of those who honestly thought, because someone taught them this: "Women can't think," and "niggers can't faint."
>
> (Walker 2010, 186; cf. Singer 2009)

But if such points had been all that the essay was about, it would not have become as famous as it did (others had made similar points before) and would not have been assigned to students in composition classes as a fine piece of writing ("En 11 Composition" n.d.). What makes it such an exceptional piece of writing is that intricately woven into the text is a moving autobiographical story with which Walker seeks to illustrate her arguments, a story of her relationship with a horse named Blue and with horses in general. The intricacy of this rhetorical device consists in Walker's story doing more than merely putting in concrete terms the parallels between the oppression of animals and the oppression of African Americans and various other minority groups. It also aims to show some parallels between the individual experience of the horse and the experience of the author herself, the parallels that are signaled by the very title of the piece, which can refer both to the author's emotional state and to her identification with the animal, who is named Blue, after all.

The essay begins with a narrative exposition depicting the period in Walker's life when she lived with her partner and her partner's son in a house in the country. Adjacent to their house was the property of their neighbors where the latter boarded Blue (Walker 2010, 183). The horse was generally well kept, but as he spent most of his days alone and in the same way (just grazing), he apparently suffered immensely from boredom. Walker then describes how she tried to ease that sorry state by interacting with the horse, feeding him apples, and how that evoked in her the childhood memories of horse riding, which she had enjoyed until an accident that made her and her mother decide that kind of activity was too dangerous.

Having shared those memories with the readers, Walker then shares with us her puzzlement at the glaring contrast between how children are taught to approach nonhuman animals and how they behave toward those species when they grow older. A horse, a dog, or a cat is usually presented to children as a fellow creature that they should care for and can develop a meaningful relationship with. The child is to try to understand, and bond with, the animal and it usually does so. But somehow in their adult life, many people forget about that experience and act as if that understanding and fellowship belonged to fairytales. The same pattern, Walker argues in her essay, can be observed in how white people sometimes approach black people. She refers in particular to those numerous Americans who had been raised by black nannies and had remained with them in as close a relationship as there can be between a child and adult, but then, after they grew older, somehow treated African Americans as belonging to a category of people with whom they could not establish close relationships at all.

These observations are then followed by a description of an incident thanks to which Blue ceases to be blue, at least for a while. One day, unexpectedly, there appears on the property of her neighbors another horse, a beautiful mare whom the author calls Brown. After the initial period of getting to know each other, the two horses eventually form a relationship which seems to make them both happy. But one day, just as unexpectedly as she arrived, Brown disappears:

> Blue was like a crazed person. Blue *was*, to me, a crazed person. He galloped furiously, as if he were being ridden, around and around his five beautiful acres. He whinnied until he couldn't. He tore at the ground with his hooves. He butted himself against his single shade tree. He looked always and always toward the road down which his partner had gone.
>
> (Walker 2010, 186)

In a single masterful stroke, Walker turns what might seem to an informed observer an irrational outburst of a beast, into an expression of utmost, deepest despair, the kind that you, and me, and she could feel having lost our loved one. "If I had been born into slavery," she writes, "and my partner had been sold or killed, my eyes would have looked like that" (185).

As she learns, the masters of Blue simply chose to let Brown be impregnated by him, and once that goal was achieved, the two were immediately separated without the slightest concern for their emotional life. This, and the remark by Walker's friend, upon seeing the solitary Blue trotting alone in the field, that horses are the epitome of freedom, makes Walker conclude with a general accusation aimed at humans. She argues that all the happy cows smiling from boxes of milk, all the chickens enjoying themselves in the commercials of poultry industry, are examples of sickening hypocrisy. One day, while eating steaks, this hits her particularly strongly. "I am eating misery," it dawns upon her, and she spits out the meat (186).

The Experiment and Results (and a Confession)

While the argumentative and narrative threads of the essay are masterfully intermingled, we found a way to disentangle them and to create a version of the text that was free from explicitly articulated points, consisting solely of the story of the horse and the story of Walker's life. One thing that we wanted to do in our experiment was to compare the attitudinal impact of that stripped down variant of the text with the original version. But since it was clear from the text that Walker believed that the stories she told in the essay conveyed roughly the same message as her arguments did, we also thought that it might be worthwhile comparing the narrative-only version with the *arguments* extracted from the original text on their own as well. Again, although these arguments were scattered throughout the essay, putting them together into the shape of one argumentative text did not demand extraordinary contortions. As a result we obtained an argument-only version of "Am I Blue?" and we wanted to see whether that text could have a greater impact on attitudes than the stories with which Walker wanted to illustrate these arguments.

The main reason why we were interested in that latter question is that one of the key debates in the ethics of literature concerns precisely what is a more powerful instrument of moral change, arguments or stories. According to the philosopher Richard Rorty, for instance, arguments are at best a handy way to summarize, and therefore *entrench*, our basic moral attitudes, and in order to actually *change* those attitudes it is more effective to use stories (Rorty 1989, 2007). There are other philosophers who subscribe to such a view, and it might be again supported to an extent by psychological studies. It has been shown, for instance, that at least for some people, narrative texts about affirmative action can lead to more positive attitudes toward that social policy than argumentative texts about that same topic (Mazzocco et al. 2010), and that narrative communications are better than non-narrative communications at impacting health-related social norms (Moran et al. 2013).

To reiterate, then, in our experiment, we wanted to see (a) if stories might be better than arguments at influencing attitudes toward animals and (b) if there might be any difference in the influence exerted by narrative texts that contain arguments and those that do not. To this end, we compared the impact of three texts: Walker's original essay "Am I Blue?," a manipulated version of that text which consisted solely of the arguments contained in the original essay, and a manipulated version which comprised only the narrative parts of the original text. The study involved 220 participants (158 women), aged 19–26, recruited from four different institutions of higher education based in the same city (Wrocław).[1] The subjects were randomly divided into four groups: three groups reading one or another of the experimental texts, respectively, and one group reading the control narrative (which again was the story about the Higgs' boson particle). As in the previous studies, the participants were asked to read the texts and fill out our questionnaire. Then it was again our job to analyze their answers.

Unfortunately, all the effort they and we had put into the study yielded results that were not spectacular. That is, while consistent with our previous results, female participants in general showed more pro-animal welfare attitudes than men ($p < 0.04$, $\eta^2 = 0.02$), the attitudinal impact exerted by our experimental texts was not statistically different from the impact exerted by our control narrative ($p = 0.7$, $\eta^2 = 0.01$). In other words, neither the original essay, nor the argument-only version, nor the narrative-only version texts had any significant attitudinal impact at all.

We could only hypothesize why this was the case. One way to interpret these negative results was as casting doubt on our previous, positive results. Perhaps, although statistically significant, our positive results were still due to chance or error, while the results of the Walker study were not marred by such problems and correctly showed that narratives do not influence attitudes toward animal welfare. This was of course possible. But note that in the case of the Walker study, the argumentative text did not work either. This suggests that our results in this study might have been due to a confounding factor that was *not* related specifically to narrativity.

Having pondered this for a while we eventually found a possible suspect: a certain fact which we had neglected when choosing our textual material for this experiment. Namely, the fact that all our experimental texts were written from a subject position that was alien, and perhaps even alienating, to our subjects. Although we did not control for this, according to our confederates none of the participants in our experiments were of African descent, and, according to the available statistical data on Polish higher education, practically all of them would have been ethnically Polish (Siwińska n.d.). Note also that most likely, the majority of these ethnically Polish participants were born and spent most of their lives in our homeland, which is important because Poland is one of the world's most ethnically and racially monolithic countries ("The World Factbook" 2017), something which is reflected precisely in the fact that we simply did not have to control for race or ethnicity. This is not a common practice in Poland, and justifiably so. Perhaps, then, asking our participants to read a text written in a very personal tone by an African-American author and accusing white society of injustice was a methodological error? We cannot be sure of this, just as we unfortunately cannot be sure whether the negative result was due to some other confounding factor or to no confounding factor at all.

Dostoyevsky's *Crime and Punishment,* or "Just the facts, Ma'am!"

How Stories Suggest What to Think, or Between Pokémon and Harry Potter

It is safe to hypothesize that when somebody tells you a story, in most cases you can easily know how that person feels about the events described. Sometimes the teller will simply inform you about it by saying, "For God's sake, this was horrible…" or "I was relieved to learn that…" or "How could she do that to me?" But very often evaluative claims about events related in a story are conveyed in a less explicit manner; not through commentary but through the mere choice of adjectives, adverbs, and verbs, through the tone of voice, through gestures and allusions (Toolan 2013, 56).

For instance, it suffices to preface the statement "People in the West have been omnivores for ages" with the word "unfortunately" to make clear the indictment. And it makes a huge difference whether somebody chooses to say "In 2013, the meat and poultry industry processed 8.6 billion chickens" (Brunker and White 2015) or, rather, "In 2013 alone, 8.6 billion innocent chickens were murdered by the meat and poultry industry." The difference we have in mind here is that the former statement is simply descriptive rather than evaluative as well. It describes a particular fact without conveying a judgment on whether that fact is good or bad. It is normatively neutral at least in the sense that it could be used in the same sense by a person who judged the fact approvingly, disapprovingly, or not at all. The statement which uses the verb "murder," in turn, does indeed convey a moral judgment, or evaluation, even if it does not do so explicitly.

More than that, aside from conveying a judgment, that statement might also make you disposed to make that judgment yourself. At least this is suggested by the large body of research showing that verbal hints such as emotionally or normatively loaded vocabulary can prime people to adopt a particular stance toward all sorts of things, from political candidates to Pokémon. What priming means here, basically, is a process where verbal cues activate certain notions, feelings, attitudes, or data in your memory that then influence how you perceive what comes next in the chain of your experience. To give two examples, there is a study which shows that "the mere co-occurrence of a fictitious character (i.e., a Pokémon) with negative words or pictures leads to a negative reaction toward this fictitious character" (Crano and Prislin 2008, 88), while another study found that

subjects subliminally primed with the verb 'to trust' ... or with
closely related verbs (e.g., to approve, to accept, to agree) ... not only
evaluated more positively the message [presented to them by the ex-
perimenters] and its source, but also expressed more intentions to
act in line with the message.

(Légal et al. 2012, 359–60)

Needless to say, all of the aforementioned ways of conveying a judgment
can be found in literature as well (Hogan 2003, chap. 2). Whether they
are willing to acknowledge this or not, all authors have certain attitudes
toward the events they describe, and they often express them in their
works. As we have seen in the case of "Am I Blue?," authors may do so in
an explicit manner, by letting their judgments and opinions be expressed
by a narrator with whom they can be clearly identified, or by prefacing
their work with an appropriate commentary.

But very often they want to be suggestive rather than explicit. And so,
for instance, they manipulate the vocabulary they use to spin their narrative, or they let their opinions be expressed by certain characters in the
story (preferably those whose features make them disposed for that purpose, positive ones in particular). Recall the example of how the *Harry
Potter* book series was criticized in Poland by a conservative scholar
for expressing a liberal ideology (Suchecka and Szpunar 2016). Leaving
aside where we stand in the conservative versus liberal debate, we have
to admit that the scholar was on to something as it has been shown empirically that the *Harry Potter* novels actually promote liberal attitudes
at least in the sense that their readers appear to be less prejudiced against
out-groups than people with the same demographic characteristics who
have not read them (Vezzali et al. 2015).

But how did J.K. Rowling achieve that? Certainly, the books themselves are *not* prefaced by Rowling's fiery sermons against prejudice, nor
are such sermons given by the narrator, nor does the narrator make a
case against prejudice in any other explicit way. But such a case is still
frequently made by the series' positive protagonists, including Harry
Potter himself. In other words, the book *does* have a certain political or
moral agenda, it *does* suggest adopting certain attitudes that are consistent with that agenda, and apparently it succeeds.

Couldn't we expect that similar instruments might also help animal
stories change attitudes toward animals? Perhaps stories of harm to animals would exert more attitudinal influence if it was *suggested* to the
reader how he or she should feel about it rather than if the suffering is
merely described? We will not hide here the fact that we had been led to
ask these questions by a specific text, one that many of our readers may
know too. It is the scene of horse beating from Fyodor Dostoyevsky's
Crime and Punishment. We had always thought it to be a very powerful
indictment of cruelty against animals. But it had never escaped us too

that this powerful indictment is not made directly, but with the help of the novel's main protagonist, or rather one of his former selves. It is this text that we used in an experiment that was meant to answer some of the above questions.

Crime and Punishment, *or a Murderer and His Horse*

Crime and Punishment is one of those rare works of literature which might be said to have endured the test of time, and can therefore (at least according to Samuel Johnson's famous doctrine, or David Hume's theories) be considered truly great.[2] The book has been celebrated by scholars and writers ever since it was published and is still widely read by lay readers – not only those who are forced to do so by their teachers but also by those who read it by choice in their leisure time. Perhaps the word "leisure," though, is not entirely appropriate here as Dostoyevsky's novel is one of the darkest and cruelest of all the books that meet Johnson's criterion. There is definitely something odd in saying that one *enjoys* this tale of a Russian student named Raskolnikov who is led by his sorry condition and Nietzschean doctrines to commit a double axe-murder (Stellino 2015, 190–95); one which he planned, and another which he did not.

The victim of the former was Alyona Ivanovna, a pawnbroker whom Raskolnikov wanted to rob. The other victim was an accidental witness, Alyona's half-sister Lizaveta, an adult with the mind of a little girl – good-hearted, naïve, and helpless: "a complete slave to her sister, [who] worked for her day and night, trembled before her, and even suffered her beatings" (Dostoyevsky 1993, 61). Raskolnikov not only did not plan to hurt her but planned the murder of the pawnbroker in such a way as to avoid Lizaveta being in the apartment when it would take place. But she happened to return to the apartment just in time to find the bloodied Raskolnikov searching the treasures of her dead sister. Anybody who has been ever engrossed by this novel, will be forever haunted by what happens next:

> He rushed at her with the axe; she twisted her lips pitifully, as very small children do when they begin to be afraid of something, stare at the thing that frightens them, and are on the point of crying out. And this wretched Lizaveta was so simple, so downtrodden, and so permanently frightened that she did not even raise a hand to protect her face, though it would have been the most necessary and natural gesture at that moment, because the axe was raised directly over her face. She brought her free left hand up very slightly, nowhere near her face, and slowly stretched it out towards him as if to keep him away. The blow landed directly on the skull, with the sharp edge, and immediately split the whole upper part of the forehead, almost

to the crown. She collapsed. Raskolnikov, utterly at a loss, snatched up her bundle, dropped it again, and ran to the entryway.

(79)

Not only is the reader haunted, but so is Raskolnikov, who despite his Nietzschean dreams of power is as emotionally fragile as the real Nietzsche was. One way in which in his fragility manifests itself is Raskolnikov's famous nightmare, which takes place shortly before the murder, and which involves a young Raskolnikov witnessing, alongside his father, the scene of a horse being beaten to death by a drunken crowd. That scene shocked our sensitivity when we first read it as teenagers (the novel being compulsory reading when we were high school students, just as it is today) and it retains its power even now:

> He always liked watching those huge horses, longmaned and thick-legged, moving calmly, at a measured pace, pulling some whole mountain behind them without the least strain, as if the load made it even easier for them. But now, strangely, to such a big cart a small, skinny, grayish peasant nag had been harnessed. ... Then suddenly it gets very noisy: out of the tavern, with shouting, singing, and balalaikas, come some big peasants, drunk as can be, in red and blue shirts, with their coats thrown over their shoulders. "Get in, get in, everybody!" shouts one of them, still a young man, with a fat neck and a beefy face, red as a carrot.
>
> ...
>
> The mare of course cannot pull the cart, which enrages Mikolka, who first whips her mercilessly himself, and then is helped by others, accompanied by laughter and cheering
>
> ...
>
> Several fellows, also red and drunk, seize whatever they can find—whips, sticks, the shaft—and run to the dying mare.
>
> Mikolka plants himself at her side and starts beating her pointlessly on the back with the crowbar. The nag stretches out her muzzle, heaves a deep sigh, and dies.
>
> But the poor boy is beside himself. With a shout he tears through the crowd to the gray horse, throws his arms around her dead, bleeding muzzle, and kisses it, kisses her eyes and mouth . . . Then he suddenly jumps up and in a frenzy flies at Mikolka with his little fists. At this moment his father, who has been chasing after him all the while, finally seizes him and carries him out of the crowd.
>
> ... "Papa! What did they . . . kill . . . the poor horse for!" he sobs, but his breath fails, and the words burst like cries from his straining chest.

(Dostoyevsky 1993, 56–58)

To better understand why we found this scene interesting recall the fragment of Eisnitz's *Slaughterhouse* that we used in the study described in Chapter 3. In case our quotes from that text did not show this clearly enough, we should stress that it depicts the plight of horses in a rather indifferent, matter-of-factual way. Neither Eisnitz nor her interlocutor Parrish explicitly pity the horses nor seem to judge what is done to them as morally wrong. Parrish in particular seems so callous and detached that even when he uses a term that normally expresses a moral judgment ("cruelty"), in his mouth it functions in purely descriptive terms. It sounds as if it were an admission to the effect that "I realize that the thing that I did to these animals would be described by most people as 'cruel', but I am indifferent about it myself."

In stark contrast to this kind of approach, the young Raskolnikov, as we have just seen, abhors the treatment of the mare to such an extent, and pities her so much, that he experiences an emotional collapse. The way we saw it, the figure of the child acted in the story as a kind of moral amplifier suggesting to the readers that what is done to the mare is wrong and they should feel compassion for the animal.

The Experiment and Results, or What Happens When You Try Too Hard

In our experiment, we wanted to see if Dostoyevsky's narrative would work differently if it were told *without* the aforementioned normative hints. We hypothesized that it would in that its impact on attitudes would then be smaller than that exerted by the original story (assuming that the latter would exert any such impact). To see if this is actually the case we manipulated the original text by removing the child character altogether as practically all of his actions and utterances conveyed some sort of normative hints. We also extracted the few expressions of moral condemnation expressed by other bystanders in the story and some terms used by the narrator that might be perceived as critical of the perpetrators.

What was left was a near exclusively descriptive story of a drunken crowd beating a horse to death. Along with the original narrative, it would be the second of the *three* experimental texts we employed in our study. The third of these was suggested to us by the parallel drawn in the novel between the innocent Lizaveta and the innocent horse. It was the scene of Raskolnikov's murder of the sisters. We wanted to see if the killing of an innocent human being might affect the subjects' sensitivity to the suffering of innocents in general in such a way that this would translate also into their concern for the welfare of animals. As usual, our control narrative was the story about Higgs boson particles.

The experimental procedure looked exactly the same as in the previous study, and the results were almost as surprising. The original story of horse beating, which seemed to us far more moving than our variant of the text, did not exert any influence on attitudes, while the stripped down variant did ($p = 0.04$). How could that be the case? Was our hypothesis of the influence of normative suggestions in animal narrative wrong, then?

We eventually came to the conclusion that it did not have to be entirely wrong after all. Perhaps the problem with the original version of Raskolnikov's horse story was that it was simply *too* suggestive in moral and emotional terms, meaning that it was sufficiently suggestive for the readers to perceive it as having a *manipulative* intent. As a result they might have perceived it more skeptically and become less susceptible to the attitudinal impact of the horse's plight than the participants reading the descriptive variant. At least this much is suggested by the results of interesting studies in the psychology of marketing on the moderating impact of perceived manipulative intent in narrative ads. Such ads are of great interest to psychologists of marketing because of their persuasive power. As Wentzel and colleagues explain in their 2010 study:

> narrative ads are evaluated more favorably than expository, factual ads since they have a similar structure to information acquired in daily life and prompt a narrative form of processing .... Narrative processing, in turn, can enhance persuasion by eliciting strong affective reactions ... and by encouraging consumers to connect the advertised brand to themselves .... Expository ads, on the other hand, usually elicit a more analytical form of processing in which consumers engage in a logical and piecemeal evaluation of the ad's arguments .... Hence, the persuasive advantage of narrative ads can be attributed to their ability to trigger a narrative form of thinking.
>
> (511)

But Wentzel and colleagues also observed, and confirmed through experiments, that when the manipulative intent of narrative advertisements becomes "salient" (for instance due to "excessive use of attention-grabbing and emotional ... tactics"), the advantage of narrativity "disappears," apparently because this kind of perceived intent raises suspicions in the audience, making them switch to their analytic processing mode as a result (511). Perhaps the emotional device used by Dostoyevsky (the sobbing and screaming of an innocent child) was indeed perceived by readers as excessive, and hence the surprising result we obtained?

While we could not draw a definite conclusion about this particular aspect of the study, there were still some firm lessons that we could draw from it. One was that here we had another experiment which showed that a text depicting animal suffering changed attitudes toward animals. Another is related to the fact that, as turned out in our analyses, reading about the axe murder of Lizaveta and the pawnbroker led to no observable improvement of the subjects' attitudes toward animal welfare. This suggested that the positive results obtained in our previous studies had not been merely due to the fact that our narratives portrayed unnecessary suffering as such, independently of whether the sufferer was human or not. It turned out, then, that the fact that the sufferers depicted in those narratives were nonhuman animals was important too.

Saunders's *Beautiful Joe*: First person vs. Third Person

What Is First-Person Narration, and Why It Matters

In the introduction to this chapter, we put forward the hypothesis that the impact of *Black Beauty* and *Beautiful Joe* might have been dependent on the first-person narrative techniques they employed, and we promised to provide experimental data indicating if that hypothesis is sound. But before we do that we should again recall the danger of chasing a chimera discussed in Chapter 1, and ask ourselves what we actually mean by "first-person narration." This is necessary as, unfortunately, distinguishing types of narrators according to the criterion of persons is not as simple as it seems, or as it is taught at schools for that matter. It simply does not suffice to say that we have first-person, second-person, and third-person narratives. Already 50 years ago, the prominent literary critic Wayne Booth (1983, chap. 6) undermined such a crude distinction with problematic cases that were so numerous that they made the very distinction itself seem problematic and dubious. But while the typology of narrators he himself provided is surely useful and sophisticated, as are many other typologies which have been put forward over the years (Birke and Köppe 2015), we will confine ourselves to stipulations that are absolutely necessary for our project, leaving all other niceties aside.

What we are concerned with in this chapter is not *any* first-person narrator, but a first-person *dramatized* narrator, who is also the main protagonist of the story he or she tells (Booth 1983, 153). What do we mean by that? One useful way to answer that question would be to ask another question: What have *we* just done? Of course, *we* asked a question. But who is this "we" anyway? It is obviously us, the authors of this book, the people who conducted the experiments described in it. But note that the "we" in question is also the *narrator* of that story. The present book is in fact a first-person narrative (that is, first-person plural narrative,) whose narrator is also its main protagonist. *We* tell a story about what *we* did, the same way the main protagonist, and at the same time the narrator, of Melville's *Moby Dick* Ishmael tells a story about what he did, and the same way many other literary characters who are also narrators do.

The autobiographical narration that you find in *Black Beauty* and in *Beautiful Joe* is precisely of this kind, and from now on we will simply call it "first-person narration" or "first-person narrative voice." In what follows, we will try to see if a story told in this way could have a larger impact than the same story told by a third-person narrator. And we mean by "third-person narrator" a narrator who is *un*dramatized, instead of partaking in the story him or herself; a narrator who only *relates* the events and does not partake in them. Of course, such narrators also come in many kinds, but, fortunately, for our purposes this general definition should suffice.

Having explained these categories, we can return to our hypothesis that the impact of *Black Beauty* and *Beautiful Joe* might have been dependent on these novels' utilizing first-person narration. This hypothesis is encouraged by two kinds of evidence, historical and experimental. As for the former, note that the novels which Lynn Hunt and Steven Pinker refer to as having paved the way for human rights culture adopted similar narrative techniques (Hunt 2007; Pinker 2011). They were epistolary novels and a large part of the fictional letters they consisted of were attributed to the protagonists from the out-groups to whose emancipation the novels apparently contributed. They then definitely contained first-person narratives, and according to Hunt and other scholars this is precisely what contributed to the intense experiences of perspective-taking on the part of their readers and to the extension of their empathy to those for whom they had previously not shown much concern (Hunt 2007).

According to available experimental data, this explanation is probably right. In their 2012 study, Geoff Kaufman and Lisa Libby observed that a narrative written in first-person voice caused its readers to simulate the experience of the protagonist to a significantly greater extent than the same narrative written in third-person voice. The former narrative also had a greater impact on pro-social behavior consistent with the message of the story. As Kaufman and Libby argue, first person narratives are "more conducive to experience-taking than … third person narratives by virtue of creating a more immediate sense of closeness and familiarity to the main character." Third-person stories, in turn, "explicitly position protagonists as separate entities (and, in our view, are more likely to position readers as spectators)" (3). This is where the difference in impact apparently comes from.[3]

Although they do not do so in their paper, in supporting that explanation Kaufman and Libby could have invoked the interesting data obtained four years earlier by Daniel Ames and colleagues (2008) which indicates that narrative voice influences the activity of the ventromedial prefrontal cortex, a part of the brain which has been shown to be activated when people think about themselves. In their study, Ames and colleagues showed their participants the faces of two unfamiliar people, asking them to imagine how these people might experience "a common event, such as 'meeting a friend for lunch'" (642). In the case of one of these people, the participants were explicitly asked to adopt that individual's perspective and write a first-person narrative describing that individual's experiences of the event in question. In the case of the other, they were asked to adopt a spectator perspective: to think about the relevant features of that person that might help to reconstruct his or her experience and then to relate that reconstructed experience in a third-person story.

Then, the experimenters asked the participants to fill in a questionnaire which included items about the preferences of each of the persons they wrote about (e.g. how much they like playing video games), with each item being accompanied by the photograph of the person the item concerned. But the participants did not simply complete the questionnaires. They completed them while having their brains scanned with functional magnetic resonance imaging technology. This technology showed that when they filled out the items about the person they previously wrote about in first-person voice, the ventromedial prefrontal cortex was activated to a significantly greater extent than when they completed items about the other individual. In other words, they thought about that individual utilizing to a greater extent the part of their brains that is responsible for thinking about themselves. What these results suggest, argue Ames and colleagues, is that

> conscious attempts to adopt another person's perspective may prompt perceivers to consider that person via cognitive processes typically reserved for introspection about the self. Consistent with earlier proposals regarding the mechanisms underlying perspective taking ..., our results suggest that the prosocial effects of perspective taking, such as increased empathy and reduced prejudice, may result from a blurring of the distinction between self and other.
>
> (643)

These are encouraging results indeed, but neither of the experiments described above, nor any other experiment on the psychological impact of first-person versus third-person narratives that we know of, concerned animal stories, which leaves a gap in our knowledge that requires filling. Note that we are not trying here to make the argument from the importance of filling a lacuna for the sake of completeness, which is unfortunately all-too-well-known in scholarly literature and makes a mockery of scholarship and scholarly progress. The gap in our knowledge that we are concerned with is *independently* interesting in that there are real obstacles to believing in a nonhuman animal telling his or her life story. After all, animals are not known to tell stories and therefore reading a story told by an animal demands quite an unusual stretch of imagination, bordering on belief in the fantastic (Cosslett 2006, 1).[4] Of course, many people do read stories precisely in order to stretch their imagination to such an extent, but we can be sure that this is not something common to all readers. Many prefer texts that stick to common sense, describing plausible sets of events, and so they shun fantastical plots. Therefore, despite the general persuasiveness of first-person stories generally, the influence of first-person animal narratives on attitudes may be limited.[5] In order to answer that worry, we had decided to ask our subjects to read an animal story in two variants, a first-person and a third-person one, and compare the attitudinal impact of the two variants between each other and with our control narrative.

From *Black Beauty* to *Beautiful Joe*

Our choice of the experimental text was a fragment taken from *Beautiful Joe* by the author Marshall Saunders (2015), a book which has been called the Canadian equivalent of *Black Beauty*. One reason it is described this way is because of the similarities in form and content, some of which it wears already on its sleeve, or in its title anyway. As in *Black Beauty*, we have an animal (this time a dog) who having suffered unspeakable cruelty at the hands of some humans found sanctuary in the home of others. Again, the animal is given a voice and speaks to the reader directly, and again the words attributed to it are ostensibly crafted so as to appeal to readers' compassion and move them deeply. But the novel is called the Canadian *Black Beauty* also because of its popularity and impact. It was the first book coming from that country "to sell over a million copies," a result it achieved in large part due to its immediate international popularity (Gerson 2010, 98; cf. Davis 2016, 242). In addition to that, and importantly for us, it is also said to have "defined the international movement that changed the way people treat animals" (Chez 2015).

However, there are some differences between the novels, the most important of them being that while *Black Beauty* was an entirely fictional character, Beautiful Joe was a real animal. Equally real was also his plight which the novel narrates. It begins almost at Beautiful Joe's birth. The readers learn about the animal's mother, siblings, and how he forms bonds with them despite the miserable conditions they are all kept in by their master, a repulsive milkman, cruel not only to his dogs, but also to his cows and horses. One dramatic day, the milkman's cruelty erupts in a bloody feast in which all of the siblings of Beautiful Joe are killed, and eventually leads to his mother's depression and death. Subsequently enraged, Beautiful Joe attempts to take revenge and is punished by having his ears and tail cut with an axe. Fortunately, the horror is witnessed by a passer-by, who rescues Beautiful Joe from the milkman and takes the dog home where he finds love and understanding. Although this sounds very much like a happy ending, it is only the beginning the adventures described in the book, which fill out 34 more of its chapters.

The Experiment and Results

The reason why in our synopsis above we focused on the initial part of the novel is because it was the source from which our experimental narrative was drawn. We used the passages describing the dog's early life in the shed owned by the cruel milkman all the way to the day when the dog was mutilated and rescued. Additionally, we modernized the vocabulary of the available Polish translation and made some further edits in order to make the length of the story similar to the length of the experimental narratives we used in other experiments. This extract constituted one of our two experimental narratives. The other was a variant of that text rewritten in third-person voice. Our control narrative was the Higgs boson story.

The participants in our experiment were 174 high school students (105 women) aged 18–19. We had randomly divided our subjects into three groups (the first- and third-person condition, and the control group), asked them to read the texts and fill out our questionnaire. While consistent with our previous results, female participants in general showed more pro-animal welfare attitudes than men, and so did pet-owners ($p < 0.001$, $\eta^2 = 0.09$), there was no difference in attitudinal impact between the two versions ($p = 0.3$, $\eta^2 = 0.01$), and the impact they exerted was not statistically different from the impact exerted by our control narrative.

This time we could not immediately see any possible confounding factor that might have skewed the results, as was the case with Walker's "Am I Blue?" We then had to take them at face value, as confirming a negative hypothesis, that narratives of animal suffering, generally speaking, do not influence attitudes toward animal welfare. The picture painted by our results were starting to become complicated. But at this point in our project, we nonetheless had a few positive results under our belts, and a few more experiments to conduct. We could still hope that the picture would become clearer once we had tested all the procedures we had planned to use. Our investigation was still on.

Notes

1 The institutions were: The University of Wrocław (journalism), The SWPS University of Social Sciences and Humanities (journalism), The University School of Physical Education and Wrocław University of Science and Technology.

2 See Abrams (1953), Introduction, cf. Hopkins (2008) and Małecki (2008).

3 Note, however, the extent to which a reader of a third-person narrative adopts a spectatorial stance would also depend on the kind of the third-person narrative in question. Some third-person narratives adopt a so-called omniscient perspective, where the narrator knows much more than the

characters about the fictional world they inhabit (perhaps even everything there is to know about it), while some others adopt a perspective where the narrator's knowledge is limited, roughly, to what is known by the protagonists or even only one of them. Such a protagonist then becomes the focalizer of the narrative, a device which is perhaps best characterized by likening it to those TV reports and documentary films where the camera always follows a particular person. But still, even in such cases, we do not look at the world through the eyes of the character per se, but rather look at him or her looking at the world, which means that we remain spectators nonetheless (Farner 2014; Shen and Sussman 2003).

4 Still another question is whether it is right to use such anthropomorphizing techniques given their presenting a false picture of animal mental life and behavior (Mitchell, Thompson, and Miles 1997).

5 However, see, e.g. Tam, Lee, and Chao (2013).

Works Cited

Abrams, M. H. 1953. *The Mirror and the Lamp: Romantic Theory and the Critical Tradition*. New York: Oxford University Press.

Ames, Daniel L., Adrianna C. Jenkins, Mahzarin R. Banaji, and Jason P. Mitchell. 2008. "Taking Another Person's Perspective Increases Self-Referential Neural Processing." *Psychological Science* 19 (7): 642–44. doi:10.1111/j.1467–9280.2008.02135.x.

Birke, Dorothee, and Tilmann Köppe, eds. 2015. *Author and Narrator: Transdisciplinary Contributions to a Narratological Debate*. Linguae & Litterae, vol. 48. Berlin; Boston, MA: De Gruyter.

Boggs, Colleen Glenney. 2013. *Animalia Americana: Animal Representations and Biopolitical Subjectivity*. Critical Perspectives on Animals: Theory, Culture, Science, and Law. New York: Columbia University Press.

Booth, Wayne C. 1983. *The Rhetoric of Fiction*. 2nd ed. Chicago, IL: University of Chicago Press.

Brunker, Mike, and Martha C. White. 2015. "The Big Bucks of Bacon: American Meat Industry By the Numbers." *NBC News*, Accessed October 26, 2015. www.nbcnews.com/business/economy/look-u-s-meat-industry-numbers-n451571.

Chez, Keridiana. 2015. "Introduction." In *Beautiful Joe*. Peterborough: Broadview Press.

Coates, Ta-Nehisi. 2015. *Between the World and Me*. First edition. New York: Spiegel & Grau.

Cosslett, Tess. 2006. *Talking Animals in British Children's Fiction, 1786–1914*. The Nineteenth Century Series. Aldershot; Burlington, VT: Ashgate.

Couch, Aaron. 2013a. "Samuel L. Jackson Reads 'Breaking Bad' Monologue: 'I Am the Danger.'" *The Hollywood Reporter*, Accessed May 6, 2013. www.hollywoodreporter.com/news/samuel-l-jackson-reads-breaking-563714.

———. 2013b. "'Breaking Bad' Fans Name 'I Am the One Who Knocks' Show's Best Line." *The Hollywood Reporter*, Accessed August 27, 2013. www.hollywoodreporter.com/live-feed/breaking-bad-fans-name-i-615149.

Crano, William D., and Radmila Prislin, eds. 2008. *Attitudes and Attitude Change*. Frontiers of Social Psychology. New York; London: Psychology Press.

Davis, Janet M. 2016. *The Gospel of Kindness: Animal Welfare and the Making of Modern America*. Oxford ; New York: Oxford University Press.

Dickstein, Morris. 2005. *A Mirror in the Roadway: Literature and the Real World*. Princeton, NJ: Princeton University Press.

Dostoyevsky, Fyodor. 1993. *Crime and Punishment*. Translated by Richard Pevear and Larissa Volokhonsky. Everyman's Library. New York: Alfred A. Knopf.

Elick, Catherine L. 2015. *Talking Animals in Children's Fiction: A Critical Study*. Jefferson, NC: McFarland & Company, Inc., Publishers.

"En 11 Composition." n.d. Accessed June 12, 2017. http://faculty.fairfield.edu/rjregan/rr11f09.html.

Farner, Geir. 2014. *Literary Fiction: The Ways We Read Narrative Literature*. New York: Bloomsbury Academic.

Foulkes, A. Peter. 1983. *Literature and Propaganda*. New Accents. London ; New York: Methuen.

Gerson, Carole. 2010. *Canadian Women in Print, 1750–1918*. Waterloo: Wilfrid Laurier University Press.

Gilligan, Vince, Karen Moore, Dave Porter, Bryan Cranston, Anna Gunn, R. J. Mitte, Aaron Paul, et al. 2009. *Breaking Bad. The Complete First Season, Disc 1 & 2 The Complete First Season, Disc 1 & 2*. Culver City, CA: Sony Pictures Home Entertainment.

Gruen, Lori. 2011. *Ethics and Animals: An Introduction*. Cambridge Applied Ethics. Cambridge; New York: Cambridge University Press.

Hogan, Patrick Colm. 2003. *The Mind and Its Stories: Narrative Universals and Human Emotion*. Studies in Emotion and Social Interaction. Second Series. New York: Cambridge University Press.

Hooker, Deborah Anne. 2005. "Reanimating the Trope of the Talking Book in Alice Walker's 'Strong Horse Tea.'" *The Southern Literary Journal* 37 (2): 81–102. doi:10.1353/slj.2005.0018.

Hopkins, David. 2008. "On Anthologies." *The Cambridge Quarterly* 37 (3): 285–304. doi:10.1093/camqtly/bfn016.

Hunt, Lynn. 2007. *Inventing Human Rights: A History*. 1st ed. New York: W.W. Norton & Co.

Kaufman, Geoff F., and Lisa K. Libby. 2012. "Changing Beliefs and Behavior through Experience-Taking." *Journal of Personality and Social Psychology* 103 (1): 1–19. doi:10.1037/a0027525.

Leech, Geoffrey N., and Mick Short. 2007. *Style in Fiction: A Linguistic Introduction to English Fictional Prose*. 2nd ed. English Language Series. New York: Pearson Longman.

Légal, Jean-Baptiste, Julien Chappé, Viviane Coiffard, and Audrey Villard-Forest. 2012. "Don't You Know That You Want to Trust Me? Subliminal Goal Priming and Persuasion." *Journal of Experimental Social Psychology* 48 (1): 358–60. doi:10.1016/j.jesp.2011.06.006.

Małecki, Wojciech. 2008. "The Bad Penny of Contingency : Literary Anthologies and the Test of Time." *Journal of Comparative Literature and Aesthetics* 31: 51–59.

Mazzocco, Philip J., Melanie C. Green, Jo A. Sasota, and Norman W. Jones. 2010. "This Story Is Not for Everyone: Transportability and Narrative Persuasion." *Social Psychological and Personality Science* 1 (4): 361–68. doi:10.1177/1948550610376600.

Mazzoni, Guido. 2017. *Theory of the Novel*. Cambridge, MA: Harvard University Press.

Mitchell, Robert W., Nicholas S. Thompson, and H. Lyn Miles. 1997. *Anthropomorphism, Anecdotes, and Animals*. Albany: State University of New York Press.

Mittell, Jason. 2015. *Complex TV: The Poetics of Contemporary Television Storytelling*. New York: New York University Press.

Moran, Meghan Bridgid, Sheila T. Murphy, Lauren Frank, and Lourdes Baezconde-Garbanati. 2013. "The Ability of Narrative Communication to Address Health-Related Social Norms." *International Review of Social Research* 3 (2): 131–49.

Ngo, Bic, and Kevin K. Kumashiro, eds. 2014. *Six Lenses for Anti-Oppressive Education: Partial Stories, Improbable Conversations*. 2nd ed. Counterpoints: Studies in the Postmodern Theory of Education, vol. 468. New York: Peter Lang.

Oatley, Keith. 2011. *Such Stuff as Dreams: The Psychology of Fiction*. Chichester; Malden, MA: Wiley-Blackwell.

Pinker, Steven. 2011. *The Better Angels of Our Nature: Why Violence Has Declined*. New York: Viking.

Rorty, Richard. 1989. *Contingency, Irony, and Solidarity*. Cambridge; New York: Cambridge University Press.

———. 2007. *Philosophy as Cultural Politics: Philosophical Papers*. Cambridge: Cambridge University Press.

Saunders, Marshall. 2015. *Beautiful Joe*. Edited by Keridiana Chez. Peterborough: Broadview Press.

Shen, Dan. 2003. "Difference Behind Similarity: Focalization in Third-Person Center-of-Consciousness and First-Person Retrospective Narration." In *Acts of Narrative*, edited by Carol Jacobs and Henry Sussman, 81–92. Stanford, CA: Stanford University Press.

Singer, Peter. 2009. *Animal Liberation: The Definitive Classic of the Animal Movement*. Updated ed., 1st Ecco pbk. ed., 1st Harper Perennial ed. New York: Ecco Book/Harper Perennial.

Siwińska, Bianka. n.d. "W Polsce Studiuje 57 119 Studentów Zagranicznych Ze 157 Krajów." *Portal Edukacyjny Perspektywy*, Accessed August 23, 2017. www.perspektywy.pl/portal/index.php?option=com_content&view=article&id=2899:w-polsce-studiuje-57–119-studentow-zagranicznych-ze-157-krajow&catid=22&Itemid=119.

Snierson, Dan. 2016. "Bryan Cranston Talks Breaking Bad's Iconic 'I Am the One Who Knocks' Scene." Entertainment Weekly. July 14, 2016. http://ew.com/article/2016/07/14/bryan-cranston-breaking-bad-one-who-knocks/.

Spadanuta, Laura. 2013. "Walter White's 'I Am the One Who Knocks' Speech as Written by Other Authors." *McSweeney's Internet Tendency*, Accessed September 23, 2013. www.mcsweeneys.net/articles/walter-whites-i-am-the-one-who-knocks-speech-as-written-by-other-authors.

Spiegel, Marjorie. 1996. *The Dreaded Comparison: Human and Animal Slavery*. Rev. and expanded ed. New York: Mirror Books.

Stellino, Paolo. 2015. *Nietzsche and Dostoevsky: On the Verge of Nihilism*. Lisbon Philosophical Studies, v. 6. Bern: Peter Lang.

Suchecka, Justyna, and Olga Szpunar. 2016. "Czym Zgrzeszył Harry Potter. Lektury Szkolne Na Indeksie - Rozmowa z Dr. Hab. Andrzejem Waśką." *Gazeta Wyborcza*, Accessed October 12, 2016. wyborcza.pl/magazyn/7,124059, 21098142,czym-zgrzeszyl-harry-potter-lektury-szkolne-na-indeksie.html.

Tam, Kim-Pong, Sau-Lai Lee, and Melody Manchi Chao. 2013. "Saving Mr. Nature: Anthropomorphism Enhances Connectedness to and Protectiveness toward Nature." *Journal of Experimental Social Psychology* 49 (3): 514–21. doi:10.1016/j.jesp.2013.02.001.

"The World Factbook." 2017. Accessed 2017. www.cia.gov/library/publications/the-world-factbook/geos/pl.html.

Toolan, Michael J. 2013. *Language in Literature: An Introduction to Stylistics.* London; New York: Routledge. http://site.ebrary.com/id/10858732.

Vezzali, Loris, Sofia Stathi, Dino Giovannini, Dora Capozza, and Elena Trifiletti. 2015. "The Greatest Magic of Harry Potter: Reducing Prejudice." *Journal of Applied Social Psychology* 45 (2): 105–21. doi:10.1111/jasp.12279.

Walker, Alice. 2010. "Am I Blue?" In *Other Nations: Animals in Modern Literature*, edited by Tom Regan and Andrew Linzey, 182–87. Waco, TX: Baylor University Press.

Wentzel, Daniel, Torsten Tomczak, and Andreas Herrmann. 2010. "The Moderating Effect of Manipulative Intent and Cognitive Resources on the Evaluation of Narrative Ads." *Psychology and Marketing* 27 (5): 510–30. doi:10.1002/mar.20341.

5 Does It Matter Who It Is About?

On Chimpanzees, Lizards, and Other Main Characters

Fifty Shades of the Protagonist

In his postscript to *Lolita*, a novel so highbrow that people often forget it is an example of crime fiction (Scaggs 2005, 213; cf. Sweeney 2016), Vladimir Nabokov famously complained about the problems he had publishing it in the USA. He recalled how one publisher expressed initial interest but demanded that Nabokov make a number of changes, including to the main characters. Lolita was to be changed from a 12-year-old nymphet into "a lad" of the same age, while Humbert Humbert, a subtle European professor born into a wealthy family, was to become "a farmer" (2000, 209). We shall leave aside the possible reasons behind these suggestions and instead encourage the reader to consider what the novel would have been like had Nabokov conceded and made the alterations. Most likely, anyone who admires the book will find such a possibility nightmarish and will be glad that the author himself thought the offer merely "amusing," eventually deciding to look for a more reasonable publisher (Nabokov 2000, 209). Humbert Humbert a farmer!? That would be an *entirely different* book, not the one we love so much, at any rate.

Now, as the reader has probably guessed, our story about this counterfactual *Lolita* has a moral in tow. That is, simply, that the extent to which our reception of a story depends on the characteristics of its protagonists can indeed be *enormous* (Oatley 2011, chap. 4). To emphasize that point let us reach outside the domain of the hypothetical and consider some *actual* cases when somebody tinkered with the characteristics of the main characters of a book, and when the results provoked negative, sometimes even furious, reactions in the audience. We are thinking here in particular about the sad lot of many film adaptations of literary stories, from *Gone with the Wind* to *The Hunger Games* (Leitch 2007). As a recent example, fans of the novel *Fifty Shades of Grey* (James 2011) were so outraged at the decision to cast the actors Charlie Hunnam and Dakota Johnson as the two main characters in the film based upon the novel that they even started an online petition! (Usmar 2017) As reported by a journalist:

> Billionaire Christian Grey, 27, [the main male protagonist of the novel] is copper-haired, grey-eyed, tall, lean, and the most gorgeous

DOI: 10.4324/9780429061424-6

man on earth, according to *Fifty Shades*; Hunnam [the actor] is ripped, but blonde, bearded, blue-eyed, 33, and British. Anastasia Steele [the main female protagonist of the novel], 21, is dark-haired, with blue eyes too big for her face, free of makeup, innocent and virginal; Johnson [the actress], 23, is blonde, with blue eyes that seem worldly, and she wears bright red lipstick.

Because of their attachment to their vision of who these characters are, fans of the *Fifty Shades* books are having a hard time visualizing these actors transforming themselves for the film.

"Very disappointed!" said a commenter identified as Nina Somerhalder, posting on the a [sic] *Fifty Shades of Grey* News page on Facebook. "He looks too old and scruffy and nowhere near attractive enough. She looks naughty. Ana is supposed to look innocent. I have read the books 6 times and have the audio book on a loop in my car, but I won't watch the film … ever!" Ella-Louise Jones put it even more succinctly: "Fantasy ruined," she said.

(Brockway 2013)

One might of course treat this case as yet another example of the supposed shallowness of popular culture, something not worth exploring. But if one scratched the surface, it would turn out that such preferences as those expressed above are implicated in serious social issues, including various class, race, and gender biases. To give a clearer, but less publicized, example of such entanglements of literary characters, consider the recent television series based on the late Ursula K. Le Guin's Earthsea novels. In the series, the main character, Ged, is portrayed by a white actor, whereas in the novels, he is explicitly described as having "red-brown skin." Why the change? We do not know for sure what the intentions of the filmmakers were. But what we do know for sure is that this was done without Le Guin's permission and that, to her, it meant that the stories she wrote were "wrecked" as a result. As she explains in a piece tellingly titled "Whitewashed Earthsea":

Most of the characters in my fantasy and far-future science fiction books are not white. They're mixed; they're rainbow. In my first big science fiction novel, *The Left Hand of Darkness*, the only person from Earth is a black man, and everybody else in the book is Inuit (or Tibetan) brown. In the two fantasy novels the miniseries is "based on," everybody is brown or copper-red or black, except the Kargish people in the East and their descendants in the Archipelago, who are white, with fair or dark hair.

My color scheme was conscious and deliberate from the start. I didn't see why everybody in science fiction had to be a honky named Bob or Joe or Bill. I didn't see why everybody in heroic fantasy had to be white (and why all the leading women had "violet eyes").

It didn't even make sense. Whites are a minority on Earth now—why wouldn't they still be either a minority, or just swallowed up in the larger colored gene pool, in the future?

(2004)

Le Guin had no qualms about calling what the creators of the TV series did an act of "whitewashing." When it is applied in the context of racial politics, that term usually signifies the portraying of non-white characters (or characters who were originally conceived as non-white) by white actors, but thanks to Le Guin's piece one realizes how it might be applied to books as well. As she reveals, the problems with her "color scheme" had begun well before the TV series and concerned already the covers of her novels. The publishers wanted to have typical covers, and a typical cover of a fantasy novel depicts the main protagonist or protagonists. But it also typically depicts a protagonist who is *white*. So they would often put a white character on the cover despite the skin color of the characters in the novel itself. As Le Guin suggests, this was all because of "a blind fear of putting a nonwhite face on the cover of a book. 'Hurts sales, hurts sales' is the mantra." To which she replies: "Yeah, so? On my books, Ged with a white face is a lie, a betrayal—a betrayal of the book, and of the potential reader" (2004).

But when the race or ethnicity of the protagonist is changed from that in the original book, this is not always an act of whitewashing. Indeed, in recent years, the reverse has often been the case. That is, a character who is white in the original text might be portrayed by an actor of color. Sometimes this is done under a policy of "color-blind casting," simply choosing the actor judged best for the role independently of any consideration of their ethnicity and the ethnicity of the character they are to play. And sometimes this is done more deliberately, precisely in order to counter the hegemony in the media of people who are white. Yet, however respectable the intentions behind such casting choices may be, they sometimes provoke as much outrage as examples of whitewashing. Notable recent examples include the casting in the 2017 movie *Spiderman: Homecoming* of Zendaya Coleman, an African American actor, as the titular superhero's girlfriend Mary Jane, and the casting in the *Harry Potter* stage play of Noma Dumezweni, again a black female actor, as Hermione Granger, who is white in the original film series (Ramaswamy 2015; Schilling 2015).

Detective fiction has received this treatment as well, for instance when the Asian American actor Lucy Liu was cast as Dr. Watson in a recent television series, *Elementary*, based on Conan Doyle's classic Sherlock Holmes stories (metrowebukmetro 2012). This casting was more controversial than the previous cases in that it meant changing the gender of the original character in addition to changing their ethnicity, and the characterization involved further significant changes as well. For instance, in *Elementary*, Dr. Watson is no longer an army veteran, nor a practicing doctor, unlike in the original stories. Writing in the name of many fans of Doyle's Holmes, Lyndsay Faye (2017) summed up their response to Liu's Dr. Watson, saying: "I'm not arguing that we're purists, because we aren't in general, but we have trepidation because frankly Watson is the heart and soul of the matter – strip him of everything he stands for, and what's left of him?"

So, as we see, some features of the protagonist may carry connotations (including political and moral ones) that are likely to influence the reception of the story they feature in. But thus far we have talked only about *human* protagonists. What about nonhuman characters? Similarly to races or group ages, certain species carry different moral and political connotations than others (summed up in the saying 'a wolf in sheep's clothing', for example),[1] so it would seem plausible that the species of the animal protagonist may be at least as important for the readers as the skin color of the human protagonist.

Imagine, for instance, that the main character in *Black Beauty* (Sewell 2012) had been a different animal; not a horse but – say – a donkey. This would not have been entirely unimaginable. Donkeys, after all, were also subject to various forms of systematic cruel treatment at the time, and they could just as easily have been imagined as relating their plight in a first-person narrative, or as making friends with humans for that matter (Bough 2011). But had Sewell chosen a beautiful black donkey for her protagonist, would she have been remembered until this day? Would her book have become a bestseller? Would there have been thousands of angry readers demanding legal action on behalf of donkeys and the like?

Or imagine that *Babe* (Noonan 2015) was about a chicken, or a turkey. Chickens and turkeys can also be cute, and can be imagined to do all the things the real Babe did. Let's face it, if people can believe in a pig herding sheep, the way Babe did in the movie, they can believe anything. But would the film have become a blockbuster? Would there have been similar stagnation in the poultry market in the year of its release as there was in sales of pork products when Babe reached the cinemas? (A. O'Connor 1995)

Or what if *The Washington Post*'s "They Die Piece By Piece" (Warrick 2001) had been about laboratory rats. Would "thousands of letters, e-mails and phone calls [have] flooded in expressing gratitude and outrage"? Would the story have "had a tremendous impact on US Congress"? ("Interview with Gail Eisnitz, Author of 'Slaughterhouse'" n.d.).

In what follows, we present experimental data which should help the readers decide about these questions, and should also, we hope, let us understand better the power of character in stories in general.

"The Dead Body and the Living Brain," or Why All Readers Are Speciesists

Stories and Biases

The first experiment described in this chapter concerns the relation between the species of the protagonist in an animal story and the impact of that story on attitudes. The main reason we decided to inquire into this matter is that the psychological research on out-groups, and on valuing the welfare of others, suggests that this kind of effect might depend on the species of the animal whose suffering is represented in the narrative, and in particular on the perceived evolutionary closeness of that species to humans. It has been shown that our perception of the suffering of *the other* may depend on the group to which we perceive that *other* to belong. And the more we perceive the other in need to be similar to us, the more we value that other's welfare (Batson et al. 1995, 1997; Batson 2011). We might then expect the attitudinal effects of narrative representations to be influenced by an anthropocentric bias. That bias would be subtle, in the sense of operating automatically and unintentionally (Fiske 2016, 303–34), and would consist in the following mechanism: the more one perceives the species of the animal protagonist to be close or similar to humans, the greater the impact of the narrative on one's concern for animals will be. Call it the speciesist spectator hypothesis.

Our study involved 479 participants divided into ten groups, nine of which were exposed to an animal narrative which was the same for each group apart from the species of the protagonist, with each of the nine groups reading about an animal of a different species. We measured the impact the narrative had on their attitudes toward animal welfare and compared the results of each group.

A Tale of a Head Transplant

The story we chose for our experiment was Oriana Fallaci's "The Dead Body and the Living Brain" (2010), originally published in 1967 in the magazine "Look" and later anthologized in the collection *Other Nations: Animals in Modern Literature*. One reason for choosing this particular text was that it concerned animal experimentation, which made it relatively easy for us to create different versions by substituting a wide variety of species for the original animal protagonist such that virtually no other elements of the text had to be altered.[2]

Fallaci's text is a journalistic report on an experiment conducted by American neurosurgeon, Case Western Reserve University professor Robert J. White. White's specialty was head and brain transplants, which he hoped could help the terminally ill (McCrone 2003). Consider, for instance, a case when one's body is overtaken by cancer, but one's brain still remains unaffected. Transplanting one's head to a healthy body could then be seen as a way of saving one's life. White did not have a chance to help any human being in this way, nor to even test his idea on humans. But he nevertheless tested it on monkeys (White et al. 1971; McCrone 2003). For this reason, he has been called "Dr. Frankenstein," while his research, along with the pictures and films portraying his disoriented, post-transplantation animal subjects, is a staple of internet articles on mad scientists and mad experiments (Summers 2014).

"The Dead Body and the Living Brain" narrates one of the earlier of such experiments. It involved a rhesus named "Libby" and consisted in severing the animal's head from her body, stripping that head from all soft tissue in such a way that there remained only a skull containing the brain, and then connecting the brain to an animal donor's circulatory system, which was to support the brain's functioning. The experiment was successful – the brain's activity appeared to be normal – and after a few hours, having gathered all the data he needed, Professor White decided to terminate it, thereby "killing the brain".

The narrative focuses on the details of the experiment, the profile of the professor, his general views, and the details of the animal subject's behavior and suffering. It traces Libby's last hours, beginning with her last meal ("orange, banana, monkey chow"), through her anaesthetization before the surgery ("It was a big needle, and Libby cried, looking at it with surprise in her eyes") to her death resulting from White's operation.

It avoids making any explicit moral judgments, though at some point the journalist asks one of the members of White's team whether the severed brain, whose functioning was artificially supported by the donor body, suffers. The reply is that the brain feels most likely like a completely paralyzed person and is aware "of the senses' absence," though it does not experience "physical pain, because all the nerves have been cut off." "Psychological pain, I don't know," explains the scientist. "I have no idea if [the brain] would feel happy or unhappy or lonely" (Fallaci 2010, 123).

Experiment and the Results, or On the Origin of Species (of the Protagonist), and Why It Matters

In our experiment, we created nine alternate versions of the text, each of them substituting for the original rhesus protagonist,[3] respectively, a chimpanzee, a pig, a parrot, a rat, a cat, a lizard, a hamster, a panda, and a hen. Our choice of species was dictated mainly by the criterion of evolutionary difference. We wanted to include species from as wide a spectrum of vertebrates as the narrative constraints allowed, which in this particular story meant mammals, birds, and a reptile. While all those animals could easily meet the constraints which the narrative imposed on its potential protagonists, we did have to manipulate a few small anatomical details. However, all remaining elements of the text were kept the same, including the name of the animal.

Our subjects were 479 students (251 women) enrolled at different programs (journalism, physiotherapy, communication, tourism, electronics, and mechanics) at three different Polish institutions of higher learning (The University of Wrocław, The Wrocław University of Science and Technology, and The University School of Physical Education), aged between 19 and 37. They were randomly assigned to one of ten groups. One of those was the control group, who would read a narrative whose topic was neutral from the point of view of our study. Each of the remaining groups would read a text with a different animal as its protagonist.

The participants were invited to laboratory spaces, informed about the ostensive purpose of the study (i.e. the relationship between the personality and worldview of readers and the way they perceive texts), and asked to fill out a questionnaire whose structure was the same as that of the questionnaires used in the studies described previously, except for one thing. In the part titled "Impressions from the text," adjoining the items about the narrative's contents, was one asking the subjects how close they thought the animal depicted in the narrative was related to humans. The subjects were to answer that item by marking on a scale representing the "evolutionary distance" between "human" and "frog" the point which they thought was occupied by the animal protagonist. Once they finished filling out the questionnaire, the study was over. The participants were thanked for their participation and dismissed.

The first thing we wanted to establish when we got our hands on the data was whether the narratives, taken together, had a positive impact on the subjects' attitudes toward animal welfare. Our statistical analyses showed that they did. That is, people expressed more pro-animal welfare attitudes after reading the texts about animals. While that effect was statistically significant, it was still rather weak, meaning that the change in attitudes was not drastic ($p = 0.01$, $\eta^2 = 0.01$). Our data also indicated that women in the whole sample expressed more pro-animal welfare attitudes than men ($p < 0.001$, $\eta^2 = 0.06$).

Once we knew that the attitudinal effect was there, we proceeded to investigate whether it was influenced by a speciesist bias, the way we hypothesized it would be. So we studied whether there would be differences between the attitudinal impact of narratives with different animal protagonists. It turned out that there were such differences and that, in particular, the two narratives which stood out as having the highest attitudinal impact, in comparison with the control group, were those with the chimpanzee ($p = 0.005$) and the parrot ($p = 0.003$).

This result supports our hypothesis, at least to some extent. After all, the chimpanzee is the species morphologically closest to humans of all those featured in our experiment (Ayala and Cela-Conde 2017;

Mitchell, Thompson, and Miles 1997a). And as for parrots, which have been called "the chimpanzees of the bird world" (Boehrer 2010, 164), they are the only species in the set that is perceived as having the capacity to talk, which is otherwise held to be a distinctively human trait. It is for this reason that they have been described as "the most human of birds" (Toft et al. 2016, 262). There is little doubt that as far as common perception is concerned, the parrot and the chimpanzee are those among the species depicted in the experimental narratives which are perceived as the most similar to humans.

But on the other hand, the high impact of these stories might be explained by a factor other than perceived evolutionary closeness, such as intelligence, for instance. Note also that the other two narratives that stood out in terms of their attitudinal impact were the variants with the panda ($p = 0.02$) and the lizard ($p = 0.02$). In the case of the panda, just like in the case of the parrot and the chimpanzee, the effect might have been alike due to a feature other than its perceived closeness to humans, such as its neotenous features, for instance (Mullan and Marvin 1999, 24–28). And the case of the lizard is a different story altogether as it prima facie refutes the speciesist spectator hypothesis. It is hard to think of a lizard as evolutionary closer to humans than, say, a cat, or any other species featured in our stories, isn't it?

It is, of course, but what we have been talking about thus far are merely *our* intuitions about how the phylogenetic distance is generally conceived and these intuitions may be wrong. In order to see whether perceived closeness really played any role in our subjects' responses to the story they read, we had to know how *they* actually conceived that closeness. And this was allowed by the data from the frog-human scale. The scale consisted of a graduated straight line whose opposite poles were "the human" and "the frog"[4] where the point immediately adjacent to the former pole indicated the closest proximity to the human and the greatest distance to the frog. The participants were asked to mark the position on the scale which they thought was occupied by the animal they read about.

They did precisely that and our subsequent analyses of their replies showed that the subjective, self-reported perception of the level of kinship between humans and the animal depicted in the narrative had a significant influence on the improvement of attitudes toward animal welfare as a whole ($p < 0.001$). In other words, the narratives featuring the species perceived as closer to humans induced pro-animal attitudes to a greater degree. The effect was present even after we excluded from our analyses the data obtained from the item about the rights of apes, which could potentially skew the results as the protagonist of one of the narratives was an ape.

There was no such effect, however, in the case of evolutionary proximity conceived in terms of contemporary biology, or objectively, so to speak. This was measured by ascribing the animals included in our narratives to different clades, that is "groups of species that all descended from one ancestral species" (Cowen 2005, 38). In the nomenclature adopted in this book "clade A," the most distant to humans, includes the lizard, the parrot, and the hen. It separated from the human evolutionary lineage ca. 292 million years ago. "Clade B," which includes the pig, the panda, and the cat, separated from our evolutionary lineage ca. 102 million years ago. The separation of "clade C," which includes the rat and the hamster, took place around 92 million years ago, whereas the chimps and the lineage leading to *Homo* separated around 7 million years ago (Hedges et al. 2015). Thus understood, the objective measure of the evolutionary, or phylogenetic, distance of the protagonist from the human species was not related to any change in attitudes toward animal welfare ($p = 25$).

But what mattered the most for our study was the subjective measure of evolutionary distance, and here we obtained positive results that had some very interesting implications. In confirming the speciesist spectator hypothesis, our data showed that even if animal representations can reduce speciesist attitudes, in the sense of making us care more for other species, this effect is mediated by those attitudes themselves. Even though the narrative we used made the participants more concerned about animals, the significance of that effect depended on the perceived evolutionary proximity of the depicted animal to humans. This shows generally how pervasive speciesism is, and specifically, that researchers interested in the social impact of animal representations should not limit their scope to the biases encoded in the content and structures of texts, but extend it to the way representations are perceived.

The Mystery of the Lizard, or What Does Neoteny Have to Do with It?

Recall, however, that while this is what our data generally indicates, there is one particular case which apparently did not fit the pattern we have just described, namely the case of the lizard. That result is puzzling, and while it might be due to an error or chance, it is at least plausible that at play here was a factor specific to that animal that trumped the perceived evolutionary distance. The hypothesis we would like to propose is related to the possibility of a combination of a certain culturally contingent fact with an innate evolutionary mechanism.

This culturally contingent fact is the peculiarity of the Polish term with which the lizard is commonly described and which we used in our experimental text. The word is "jaszczurka" and what is interesting about it from our point of view is that it ends with "-ka." The suffix "-ka" is widely used in the Polish language to signify diminution, the way the suffix '-y' is sometimes used in English and the suffix '-chen' in the German language (Boxenbaum and D'Souza 2013, 139–96). In particular, when the suffix is applied to generic names of living beings, this is often done in order to signify that the being in question is a baby or is small, and to thereby trigger neotenous associations (Genosko 2005). So, for instance, if the Polish word for mouse is "mysz," the diminutive derivate of that word is "myszka," meaning either "a small mouse" or "a baby mouse," the same way in which "doggy" can signify "a small" or "baby dog" in English, and "Hündchen" "a small" or "baby dog" in German.

However, the "-ka" in "jaszczurka" is not a diminutive suffix. The standard meaning of that word in the Polish language is simply "lizard," not a "small lizard" or "baby lizard." But while the suffix "-ka" has other functions in the Polish language than to signify diminution (e.g. it is used to form the female-gendered names of professions), there is a widespread impression that it has that function predominantly, such that the ending is taken to do so even in the cases when the actual morphological function of a particular use of it is entirely different. It is precisely because of this effect that some Polish feminists have objected to the practice of using the suffix "-ka" to coin female-gendered names of professions in the Polish language. They are concerned that through their neotenous associations such names belittle the women they refer to (Kłosińska 2009).

To come back to our experiment, the "-ka" in the generic name "jaszczurka", though de facto not deployed here as a diminutive suffix, might have nevertheless generated neotenous associations in our participants. It might have thereby triggered in them specific sympathetic responses that were not at play in the case of the other animals, and in turn made the lizard story have the unexpected effect on attitudes toward animals. As argued by Konrad Lorenz and others, it is a matter of innate, evolutionary mechanism that "we collectively feel more sympathy and affection toward those animals that possess juvenile features" (Boxenbaum and D'Souza 2013, 186). However, as far our experiment is concerned, the case of the lizard was an exception to the rule (i.e. the effect of perceived species similarity and proximity). All the pieces of our puzzle finally fell into place. We had found a plausible explanation of our mystery.

The Trouble with Abstractions, or "Animals" versus Horses and Dogs

A Dead End and a Solution

No sooner had we started toasting our success solving the mystery of the lizard than we realized that there was a much larger and more troubling puzzle ahead of us. It was related to the fact that the Fallaci study was the last of the experiments we planned to conduct in order to see whether animal narratives can influence attitudes toward animals. From then on, we planned to focus on the mechanisms of the effect, if any, and its durability. It was time for us to look back at our experiments and see what they could tell us.

Unfortunately, what they said did not allow for any clear-cut conclusions. Notice that while some of our studies had shown that stories can induce attitudinal change (the Krajewski study, the Eisnitz study, and the Fallaci study), others indicated that they cannot (the Walker, and the Saunders studies), while the results of one particular study (the one on Dostoyevsky) were ambiguous. Ambiguity, then, was also the overall picture painted by our results. We still did not know for certain whether narratives, generally speaking, can immediately influence attitudes toward animals or not.

This was precisely one of those moments of doubt and confusion that we knew sometimes befall anyone conducting an investigation, scholarly or detective. The results of our investigation were inconclusive. But we could not leave our story like this. In scientific stories, like in detective stories, a clear-cut conclusion *is* obligatory, after all. It constitutes the major goal of the story and at the same time the major reward for the reader. If we wanted to have a conclusion of this kind, we knew we had to obtain more results, and therefore we had to conduct more studies than were initially planned.

This is what we did. Having pondered our previous data, we realized that there is a way to understand attitudes toward animals that is different from the one we previously employed *yet* at the same time is relevant for measuring the phenomenon we wanted to capture. Thus far, we had focused on attitudes toward animals *in general*, where that category is understood as embracing *all* animal species except humans. This sounded reasonable, but note that *Black Beauty*, for instance, was not reported to affect attitudes toward animals in general, but toward a very specific species – the one portrayed in the book. Ditto for *Babe*, the *Washington Post* story, and the book by Eisnitz (2007).

This made us realize that the criterion of *influencing attitudes toward animals in general* that we applied to our narratives was perhaps too demanding and out of touch with social reality. Consider, for the sake of argument, that we have just proven experimentally that narratives can dramatically influence attitudes toward *any* particular species, yet none of them can influence attitudes toward animals in general. What sense would it make, then, to say that they cannot influence attitudes toward animals? That would be mere sophistry.

With this in mind, we decided that before we draw any conclusions from our previous results, including the conclusion of inconclusiveness, we should take the stories that had been observed not to exert any influence on our subjects' attitudes toward animals in general and to test whether they could nevertheless change our subjects' attitudes toward the welfare of the species described in them, namely horses and dogs. We were then to return to the original version of Raskolnikov's dream, to Walker's "Am I Blue?," and to the story of the plight of Beautiful Joe.

The Experiment and Results

In order to conduct such an experiment on attitudes toward horses and dogs, we obviously needed some new instruments. First of all, we needed two new scales, one measuring attitudes toward dogs and the other measuring attitudes toward horses. Unfortunately, we could not build them by simply rephrasing the items in the ATAW scale such that they would refer to these species alone. (This would be impossible in the case of most of the items, perhaps most glaringly in the case of the one about whaling and whales.) But since we wanted the results of this experiment to be as commensurate as possible with our previous studies, we made sure that our new scales were at least similar to the ATAW scale. They comprised the same number of items (seven) and reused two items from ATAW that proved adaptable for our purposes, namely, "Human needs should always come before the needs of animals" (where we substituted "horses," or "dogs," for "animals'), and "Basically, humans have the right to use animals as we see fit" (where we did the same).

To these two items we added five entirely new ones. In the equine variant, they read as follows: "Cruelty to horses should be punished as severely as cruelty to people"; "I personally care about the plight of horses"; "Our society does not do enough to protect horses from cruelty"; "Our society should do more to protect the welfare of horses"; and "Compared with other social problems we face today (drugs, homelessness, crime, education), the question of helping horses is completely insignificant." The canine variant had "dogs" in place of horses in these items. This way we created two separate scales comprising seven items each. We called one of them Attitudes Toward Horses Scale (or ATHS for short), and the other Attitudes Toward Dogs Scale (ATDS). A pilot test showed that the psychometric parameters of the scales were good. We could then employ them in our experiment.

But before we describe how this was done, we should add two important pieces of information. First, in constructing ATHS and ATDS, we appropriated some items from a questionnaire used by the American psychologist C. Daniel Batson and colleagues in a now-classic study on attitudes toward human out-groups (Batson et al. 1997). Second, what we also borrowed from that particular study was the whole design of our experiment, as we found it suitable for our purposes and were unable to employ the one we used previously.[5] Let us then say something about the method used in the study conducted by Batson's team.

Theirs was an experiment conducted on a group of US students, and its ostensible goal was to assess the subjects' "emotional and evaluative responses" to "a brief pilot broadcast" by "the local university station" (Batson et al. 1997, 108). The subjects were informed that the broadcast was to "involve an interview with a young woman from the Kansas City area who is experiencing the personal tragedy of AIDS" (108). The interview was then presented to the subjects, and after listening to it, they were asked to fill out a few questionnaires including one about whether they thought the broadcast was "interesting and worthwhile", and another on their "attitudes toward people with AIDS" (108).

In our study, we proceeded similarly. Our subjects were 170 students, aged 19–32, including 92 women, who studied in various programs at three different institutions of higher education. They volunteered to participate in a study whose ostensible purpose was to evaluate stories that were to be featured on a planned web portal for students. In truth, no such portal was in anyone's plans, but our confederates presented a believable story that it was, including that it was funded by an EU grant. Accordingly, all the materials distributed as part of the study bore appropriate logos so that they looked as authentic as possible.

The subjects were randomly divided into five groups, of which three read our experimental texts. One read "Am I Blue?", another Raskolnikov's dream, and still another read *Beautiful Joe*. In addition to this, there were two control groups, each reading our usual narrative placebo, the story about the Higgs boson particle (and why we needed to have two control groups will become clear in a moment). The participants in each group were informed that their opinion was extremely valuable for the editors of the portal and that it would influence future decisions about the kind of texts the portal will feature. The subjects were then asked to read one of the narratives and fill out a questionnaire.

In its first part, the questionnaire featured items about the subjects' reading experiences, followed by five items measuring their "General opinion about the text," such as, for example, "I think the text was interesting," "I think the text was well-written," and "I would post a link to the text on Facebook." Then, there followed demographical questions, and a set of items asking the subjects about their views on a particular topic the portal planned to cover. In the case of the groups reading Raskolnikov's dream and "Am I Blue?" these latter items included ATHS, while in the case of *Beautiful Joe* group, they included ATDS. In each case, they were preceded by an item "The subject of animals and their welfare is important and should be covered by the media," which we thought would further contribute to the credibility of our cover story.

As to the control groups, it should be now easier to see why we needed two of those. On the one hand, we needed people in our control sample to answer questions about dogs and horses. But on the other, methodological considerations precluded us from including both ATHS and

ATDS in the same questionnaire for the control group. The reason was that such a questionnaire would be *significantly* different in its structure from those filled out by the experimental groups, and this would make the results achieved on it incommensurate with those achieved on the other two.

This way we had to have two control groups, each of which read the text about the Higgs Boson particle, and each of which had a questionnaire that consisted of the same number of questions, including exactly the same questions about their impressions from reading the Higgs boson story and their general assessment of it. The only thing that differed were the "additional" questions they answered, with the questions about horses given to one control group and the questions about dogs given to the other.

At the last stage of the experiment, our confederates collected the questionnaires and thanked the subjects for their participation. It is worth noting here that once this happened, many students were eager to talk about the portal we had described to them. They offered their advice as to what kinds of texts it should publish and wished to discuss the topics their narratives concerned, including animal welfare. This suggests that they believed our cover story used in the experiment, and that they were convinced that their input was important so will have taken their participation in seriously. Our experiment seemed to have worked well.

It worked well also in the sense that it provided unambiguous results. *Each* of our experimental texts improved the subjects' attitudes toward the welfare of the species of the protagonists of those texts, horses or dogs. This meant that even if our previous studies had shown that the power of those texts was insufficient to immediately change attitudes toward animals *in general*, they still influenced attitudes toward certain *kinds* of animals (*Beautiful Joe*: $p = 0.01$, $\eta^2 = 0.09$; Raskolnikov's nightmare: $p = 0.02$, $\eta^2 = 0.08$; *Am I Blue?*: $p < 0.01$ $\eta^2 = 0.11$).[6] Taken together, then, the results of all of our experiments yielded a positive conclusion. Yes, stories *do* have a positive impact on the way we value the well-being of other species – in one way or another.

Notice here that given the variety of the texts and audiences we worked with, the above conclusion is quite sound and generalizable. Our stories included both journalistic reports and fiction, texts from the nineteenth century (Dostoyevsky, Saunders) and contemporary ones (Walker, Fallaci, and Krajewski), and, finally, they were penned by authors from a host of different countries, including the USA, Canada, Italy, Russia, and Poland. Our subjects were both men and women and while most of them were high school or university students, we made sure that their schools were of different profiles, and we also conducted one experiment (the Krajewski study) whose participants represented all sorts of educational backgrounds and age groups. All told, our sample included more than 3,000 people (3,094, to be exact) whose age spanned between 14 and 81. We were entitled to assume that our conclusions could be extended to the general population.

But even if we finally knew that stories do impact attitudes, some issues remained to be decided before we could present a comprehensive conclusion applicable to the real-world conditions that had motivated our study to being with. For one thing, we still needed to learn more about the psychological mechanisms that were responsible for the attitudinal change we observed. And for another, we needed to know how long the effect would last. These questions are the subject of the following chapters.

Notes

1 See, e.g. DeMello (2012, chap. 3).
2 Note that this would have been much more difficult for narratives about other topics, for instance, about factory farming or slaughter, such as "They Die Piece by Piece." After all, there is only a certain narrow group of species that could believably be their main protagonists. But the context of experimental research is one in which basically any animal could believingly fit. While lab animals are usually associated with mice, rats, bunnies, and monkeys, the public regularly hears from the media about studies conducted on a myriad of other species, from elephants to ants. In other words, we thought that a story about scientific experimentation could give us the freedom we needed.
3 We decided not to retain the original rhesus protagonist fearing that some of the participants may be unfamiliar with this species, something which might skew the results.
4 The main reason for including "frog" as the opposite pole of "human" in our scale was that we wanted to make the scale as fine-grained as possible. We hypothesized that the larger the evolutionary scope of our scale (e.g., if instead of "frog" there were "jellyfish"), the more difficult it would be for the participants to decide where exactly on the line the animal protagonist should be situated. In other words, the more coarse-grained the scale would be. So we eventually decided to use as its opposite pole an animal from a group that was evolutionarily closest to humans of all the groups that were not included in the narratives (which included mammals, birds, and reptiles). That group was amphibians, which we thought to be best epitomized by the frog.
5 Recall that the apparent purpose of our previous experiments was the relation between the subjects' personality and worldview on the one hand and their reading experiences on the other, and that it involved accordingly a questionnaire apparently measuring personality and worldviews. Obviously, items concerning a concrete species such as the dog would be too specific for such a questionnaire, and given that the species of the animal mentioned in them would match the species of the animal protagonist in a narrative the subjects would read, the actual purpose of our study would be then easy for them to guess.
6 Admittedly, we did not perform symmetrical experiments for those texts that exerted the more general kinds of influence. But this was unnecessary per se for rendering our results conclusive, and, also, we could take for granted that those texts also exerted influence on attitudes toward the

species of their protagonists. Of course, while there is a logical implication from valuing the welfare of animals more to valuing more the welfare of a given species, it may still not be psychologically the case that somebody who comes to value animals more values also a given species more. But that would be very unlikely in a case such as ours, when the change comes about as a result of reading a story about an animal of that particular species. It is just hard to believe that a story about a monkey, or a horse, or a dog, might prompt somebody to care more about animals, but not prompt her to care more about dogs, monkeys, and horses.

Works Cited

Ayala, Francisco José, and Camilo J. Cela-Conde. 2017. *Processes in Human Evolution: The Journey from Early Hominins to Neandertals and Modern Humans*. Oxford; New York: Oxford University Press.

Batson, C. Daniel. 2011. *Altruism in Humans*. Oxford; New York: Oxford University Press.

Batson, C. Daniel, Eddie Harmon-Jones, Heidi J. Imhoff, Erin C. Mitchener, Lori L. Bednar, Tricia R. Klein, Lori Highberger, and Marina Polycarpou. 1997. "Empathy and Attitudes: Can Feeling for a Member of a Stigmatized Group Improve Feelings toward the Group?" *Journal of Personality and Social Psychology* 72 (1): 105–18. doi:10.1037/0022-3514.72.1.105.

Batson, C. Daniel, Tricia R. Klein, Lori Highberger, and Laura L. Shaw. 1995. "Immorality from Empathy-Induced Altruism: When Compassion and Justice Conflict." *Journal of Personality and Social Psychology* 68 (6): 1042–54.

Boehrer, Bruce Thomas. 2010. *Parrot Culture: Our 2500-Year-Long Fascination with the World's Most Talkative Bird*. Philadelphia: University of Pennsylvania Press.

Bough, Jill. 2011. *Donkey*. Animal. London: Reaktion Books.

Boxenbaum, Harold, and Richard W. D'Souza. 2013. "Interspecies Pharmacokinetic Scaling, Biological Design and Neoteny." In *Advances in Drug Research*, edited by Bernard Testa, 139–96. London: Academic Press.

Brockway, Laurie Sue. 2013. "Fifty Shades of Grey Casting Controversy: Why Fans Are So Upset." *Everyday Health*, Accessed April 9, 2013. www.everydayhealth.com/depression/why-fans-are-so-upset-about-fifty-shades-of-grey-casting.aspx.

Cowen, Richard. 2005. *History of Life*. 4th ed. Malden, MA: Blackwell Pub.

DeMello, Margo. 2012. *Animals and Society: An Introduction to Human-Animal Studies*. New York: Columbia University Press.

Eisnitz, Gail A. 2007. *Slaughterhouse: The Shocking Story of Greed, Neglect, and Inhumane Treatment inside the U.S. Meat Industry*. Amherst, NY: Prometheus Books.

Fallaci, Oriana. 2010. "The Dead Body and the Living Brain." In *Other Nations: Animals in Modern Literature*, edited by Tom Regan and Andrew Linzey, 117–24. Waco, TX: Baylor University Press.

Faye, Lyndsay. 2017. "A Holmes Fan's Mistrust of Elementary: An Open Apology to CBS by Lyndsay Faye." Accessed June 21, 2017. www.criminalelement.com/blogs/2012/05/sherlock-holmes-fans-mistrust-elementary-an-open-apology-to-cbs-lyndsay-faye.

Fiske, Susan. 2016. *Social Cognition: From Brains to Culture.* 3rd edition. Thousand Oaks, CA: SAGE Publications.

Genosko, Gary. 2005. "Natures and Cultures of Cuteness." *InVisible Culture: An Electronic Journal for Visual Culture* 9. www.rochester.edu/in_visible_culture/Issue_9/genosko.html.

Hedges, S. Blair, Julie Marin, Michael Suleski, Madeline Paymer, and Sudhir Kumar. 2015. "Tree of Life Reveals Clock-like Speciation and Diversification." *Molecular Biology and Evolution* 32 (4): 835–45. doi:10.1093/molbev/msv037.

"Interview with Gail Eisnitz, Author of 'Slaughterhouse.'" n.d. A Vegan Skeptic. Accessed January 26, 2017. www.wegodlessanimals.com/inteview-with-gail-eisnitz-author-of-slaughterhouse/.

James, E. L. 2011. *Fifty Shades of Grey.* First Doubleday Hardcover Edition. Fifty Shades Trilogy, Book 1. New York: Doubleday, a division of Random House, Inc.

Kłosińska, Katarzyna. 2009. "Feminizm w Języku Polskim." *Polityka*, August 25, 2009. www.polityka.pl/tygodnikpolityka/spoleczenstwo/299523,1,feminizm-w-jezyku-polskim.read.

Le Guin, Ursula K. 2004. "A Whitewashed Earthsea." *Slate*, Accessed December 16, 2004. www.slate.com/articles/arts/culturebox/2004/12/a_whitewashed_earthsea.html.

Leitch, Thomas M. 2007. *Film Adaptation and Its Discontents: From Gone with the Wind to the Passion of the Christ.* Baltimore, MD: Johns Hopkins University Press.

McCrone, John. 2003. "Monkey Business." *The Lancet Neurology* 2 (12): 772. doi:10.1016/S1474-4422(03)00596-9.

metrowebukmetro. 2012. "Lucy Liu to Play Joan Watson in US Sherlock Holmes Remake Elementary." *Metro* (blog). Accessed February 28, 2012. http://metro.co.uk/2012/02/28/lucy-liu-to-play-joan-watson-in-us-sherlock-holmes-remake-elementary-333750/.

Mitchell, Robert W., Nicholas S. Thompson, and H. Lyn Miles. 1997. *Anthropomorphism, Anecdotes, and Animals.* Albany: State University of New York Press.

Mullan, Bob, and Garry Marvin. 1999. *Zoo Culture.* Urbana: University of Illinois Press.

Nabokov, Vladimir Vladimirovič. 2000. *Lolita.* London: Penguin Books.

Noonan, Chris. 2015. *Babe.* DVD.

Oatley, Keith. 2011. *Such Stuff as Dreams: The Psychology of Fiction.* Chichester; Malden, MA: Wiley-Blackwell.

O'Connor, Amy. 1995. "When Pigs Fly." *Vegetarian Times*, December, 16.

Ramaswamy, Chitra. 2015. "Can Hermione Be Black? What a Stupid Question," December 21, 2015, sec. *Stage.* www.theguardian.com/books/shortcuts/2015/dec/21/hermione-granger-black-noma-dumezwani-harry-potter-cursed-child.

Scaggs, John. 2005. *Crime Fiction.* The New Critical Idiom. London; New York: Routledge.

Schilling, Dave. 2015. "Black Hermione: Which Character Will Cause the Next 'Racebending' Outrage?" December 22, 2015, sec. *Culture.* www.theguardian.com/culture/2015/dec/22/black-hermione-racebending-white-characters-idris-elba-james-bond.

Sewell, Anna. 2012. *Black Beauty*. Oxford: Oxford University Press.

Summers, Ken. 2014. "The Shocking Experiments of Robert White: Cleveland, Ohio's Own Dr. Frankenstein." Accessed November 7, 2014. http://weekinweird. com/2014/11/07/robert-white-clevelands-dr-frankenstein/.

Sweeney, Susan Elizabeth. 2016. "Whether Judgments, Sentences, and Executions Satisfy the Moral Sense in Nabokov." In *Nabokov and the Question of Morality*, edited by Michael Rodgers and Susan Elizabeth Sweeney, 161–81. Palgrave Macmillan US. doi:10.1057/978-1-137-59221-7_10.

Toft, Catherine Ann, Timothy F Wright, James D Gilardi, and World Parrot Trust. 2016. *Parrots of the Wild: A Natural History of the World's Most Captivating Birds*. Oakland: University of California Press.

Usmar, Jo. 2017. "Fifty Shades of Grey Film: Petition to Protest Casting of Charlie Hunnam and Dakota Johnson – Mirror Online." Accessed June 14, 2017. www.mirror.co.uk/tv/tv-news/fifty-shades-grey-film-petition-2253421.

Warrick, Jo. 2001. "'They Die Piece by Piece.'" *The Washington Post*, April 10, 2001. www.washingtonpost.com/archive/politics/2001/04/10/they-die-piece-by-piece/f172dd3c-0383-49f8-b6d8-347e04b68da1/?utm_term=.63950 eaf0c71.

White, Robert J., Lee R. Wolin, Leo C. Massopust, Norman Taslitz, and Javier Verdura. 1971. "Cephalic Exchange Transplantation in the Monkey." *Surgery* 70 (1): 135–39. doi:10.5555/uri:pii:0039606071900997.

6 How Does It Work?

From Readerly Pleasure to Animal Cruelty

Howdunit (and Why?)

There is a very good reason why detective stories are often called "whodunnits." While there are significant exceptions to this rule, most novels in that genre simply *are* about the process of finding out *who* committed a crime and then catching him, or her (Herbert 2003; Rzepka and Horsley 2010, 43–45). Once this is achieved, the plot is dissolved and the book ends. This is also how most readers *want* detective novels to end. They apparently do not care what happens next with the criminal and the crime. It is almost as if the closing of handcuffs on the wrists of the perpetrator brought all consequences of their crime to a close as well. As if the boundaries of the world of detective novels ended where the arrests began. As if there were nothing beyond them.

But of course in reality, there is so much going on with the criminal and the crime after the arrest that it constitutes an entirely separate world; there is enough going on, at any rate, that it apparently merits at least one other artistic genre, the courtroom drama (Sauerberg 2016). Anyone who knows anything about that world, from novels, films, or otherwise, will also know that from the point of view of criminal justice, the question of "whodunit," while crucial, is only half of the story. For in order to judge and punish the criminal, it will also have to be established how, and why, the deed was done. Consider *Breaking Bad*'s Walter White, for example. Even if you have little sympathy for him, it is clear that his original motive for going into the methamphetamine business, and the fact that he, initially at least, tried to run it with as little violence as possible, should be taken into account by a judge or jury for their verdict and sentencing to be just.[1]

But the knowledge of how and why is essential not only for judging particular criminal acts but also for effectively dealing with the kind of crime they represent on a social scale. When taken collectively, the data on how and why particular murders were committed obviously help us understand how and why people tend to commit murders. The same applies, mutatis mutadis, to data on theft, bribery, or dog fighting,[2] and so on. Once we understand *that aspect of criminal activities*, we are in a

DOI: 10.4324/9780429061424-7

better position to prevent these kinds of crime in the future. This is the province of criminology and crime prevention units.

And if our research is much like detective investigation, then it is also very much like the judicial process and criminology. From the very beginning of our project, we were interested not only in catching our suspect but also in investigating what prompted its actions and the mechanisms by which they were performed. And what we wanted to draw eventually from that data were practical implications for animal welfare policies. In other words, we wanted to know how and why it is that animal narratives influence attitudes toward animals. And we wanted to know that so we could use it to best exploit this effect in social practice.

The existing literature on the psychology of narratives and attitude change had suggested to us to the following two factors that might be responsible for the effect we observed in our previous experiments. That is, we hypothesized that the influence of narratives on attitudes toward animals might depend to a significant degree on the extent to which those stories can absorb their readers and on the extent of cruelty described in them. Both these factors appeared to us to be fundamental here, and both of them are related to questions that have recently been the subject of considerable scholarly attention. These are the question of so-called transportation into text (Green and Brock 2000) and the question of aesthetic representations of suffering (Di Bella and Elkins 2013). In what follows, we will try to make as good a use of this exciting body of work as possible and also to make a genuinely novel contribution to it.

Lost and Found (in Stories)

Don't Read and Drive!

In January 2016, a policeman in Minnesota spotted a car on a highway that was behaving quite suspiciously. The vehicle drove unusually slowly and was swerving. Obviously, something was wrong with the driver. This might have been something dangerous but innocent, such as falling asleep. But it was also possible that something more serious, and criminal, was going on, such as the driver being drunk, or on drugs, or being involved in a fight with a passenger. It was definitely something for the police to investigate.

But when the car was pulled over, the reason turned out to be neither trivial nor something *from* a crime thriller. It was a *thriller* itself. That is, as the driver explained, the reason he drove so erratically was that he was at the same time reading a recent thriller by James Patterson. The driver apparently said he just could not put it down ("Eagan Police Ticket Man For DWR: Driving While Reading" 2017). Obviously, the officer on his part could not let the driver get away without a fine, and equally obviously, the media could not pass over that story without comment.

It was subsequently featured in all kinds of newspapers, TV channels, and all over the internet, thanks to which we can report it to you here.[3] And we do so because it is an extreme example of the phenomenon that we want to study in this subchapter, a phenomenon which is considered one of the greatest readerly pleasures and usually called being "lost in a book" or a story (Nell 1988).

All of us have been there before, absorbed by a story so much that we forgot about the entire world with its problems, worries, duties, and so on. As we have just seen, some people can get so lost in a story that they become entirely unaware of dangers that would otherwise make their hair stand on end. The case of the James Patterson reader ended well, but there has been at least one documented case of a driver actually causing a serious accident because he was reading a book while behind the wheel ("Driver Reading Behind the Wheel Causes Crash" 2017). Indeed, that he caused an accident should not surprise us as it has been shown that people who are engrossed in activities that involve high levels of visual load such as reading, tend to become genuinely out of touch with the outside world. For example, they become so insensitive to sounds that this was called "inattentional deafness"! (Macdonald and Lavie 2011)

It's the Transportation, Stupid!

The phenomenon in question is indeed fascinating and like other fascinating, but unfamiliar, effects it is usually talked about in metaphorical terms. In referring to it, we often say that we are lost in a story, that we are absorbed or consumed by it, that we are transported into another world, and the like (Green and Brock 2000). Given how widespread and essential for the experience of reading that phenomenon is, our readers may be surprised to learn that its role in the psychological effects of narratives became the subject of serious research only recently. However, once this research trend started it ballooned to impressive proportions. This explosion of interest can be traced to a single study, published in 2000, and hitherto cited more than 2,000 times.

The study, titled "The Role of Transportation in the Persuasiveness of Public Narratives" and authored by Melanie Green and Timothy C. Brock, proposed a technical term for the phenomenon of being lost in a story, its definition, as well as an instrument with which to measure it. Finally, it established experimentally that this phenomenon plays a crucial role in the persuasive power of stories. The term was "transportation"; it was defined as "a convergent process, where all mental systems and capacities become focused on events occurring in the narrative"; while the instrument for measuring it was an 11-item scale which aimed to capture the three consequences of transportation that the researchers considered the most crucial:

The first consequence ... is that parts of the world of origin become inaccessible. In other words, the reader loses access to some real-world facts in favor of accepting the narrative world that the author has created. This loss of access may occur on a physical level – a transported reader may not notice others entering the room, for example – or, more importantly, on a psychological level, a subjective distancing from reality. While the person is immersed in the story, he or she may be less aware of real-world facts that contradict assertions made in the narrative. ... Beyond loss of access to real-world facts, transported readers may experience strong emotions and motivations, even when they know the events in the story are not real .... For example, when transported into narratives with unhappy endings, transported individuals are likely to engage in what Gerrig ... termed *anomalous replotting:* "actively thinking about what could have happened to change an outcome" .... A third consequence is that people return from being transported somewhat changed by the experience.

(Green and Brock 2000, 702)

Added to the aforementioned 11 items were four items designed to measure the vividness of the imagery invoked by particular elements of the text, such as its protagonist for instance.

This instrument became popular in the psychological community to the extent it did mainly because, using it, Green and Brock were able to establish some important things about the persuasive power of stories, for instance, that absorption in a story can help reduce or eliminate what they call negative cognitive responding to a story, that is adopting a critical stance toward it. This is because "transported individuals are so absorbed in the story that they would likely be reluctant to stop and critically analyze propositions presented therein" (Green and Brock 2000, 703). In other words, stories can not only be so hard to "put down" (to quote the Minnesota driver) that you do not want to scrutinize what is going on outside of them but that you do not even want to scrutinize what is going on *within* them. Since it is impossible to critically analyze a story while being absorbed by it (as impossible as, say, critically analyzing a game of tennis while being absorbed by it), and since being absorbed by it is a source of enormous pleasure you typically want more of, you just do not want to do it, and instead go on reading.

While this feature of transportation could in itself directly boost the persuasive effects of any story, there are, according to Green and Brock some other, indirect ways in which it could do it:

transportation may make narrative experience seem more like real experience. Direct experience can be a powerful means of forming attitudes ..., and to the extent that narratives enable mimicry of experience, they may have greater impact than nonnarrative modes. Finally, transportation is likely to create strong feelings toward story characters; the experiences or beliefs of those characters may then have an enhanced influence on readers' beliefs.

(2000, 702)

Given the above, it is no wonder that Brock and Green posited that "transportation is the key determinant of narrative impact." In their paper, they further presented the results of as many as four experiments which directly or indirectly supported that claim as well as the more specific claims quoted above.

Most importantly, their results showed that those of their subjects who were more transported into an experimental story showed more "story-consistent beliefs" than those who were less transported. One of those stories, for instance, talked about the murder of a small girl committed by a psychiatric patient in an Ohio mall. Having read it, the participants were asked to say how often stabbing murders happen in Ohio malls, and the estimated frequency of such events given by those participants who found the story more absorbing was higher than that indicated by those who found it less absorbing (Green and Brock 2000, 706).

Another thing that Green and Brock's results showed was that those of their participants who were more transported showed more positive attitudes toward the positive characters in the story, and that they found fewer places in the story that rang false to them. This is how the history of empirical study of transportation in narratives began, and since then it has been shown to influence a dizzying array of things, including some that are very interesting from our point of view such as the impact which narratives have on empathic attitudes, and on attitudes toward out-groups. In his 2016 study, for instance, Dan Johnson asked his subjects to read a short story "about a counterstereotypical female Muslim protagonist who is verbally and physically assaulted in a subway station but confronts her attackers" (2016, 80). Again, those who were more transported into that story, showed more positive attitudes toward Arab-Muslims as a result.

Given the parallels between attitudes toward human out-groups and attitudes toward animals (Plous 2003), and given the similarities to which our previous results pointed in the way narratives can impact these two kinds of attitudes, we expected the attitudinal impact of animal narratives to be similarly influenced, or mediated, by transportation. Since no experimental evidence that would directly support this hypothesis was available, we had to provide it ourselves. In this case, however, it was not necessary for us to conduct a new experiment as the necessary data was waiting buried in the questionnaires completed by the subjects of our previous studies. Recall that some of those questionnaires contained items which concerned the readers' impressions from the text, and that these items included the scale devised by Green and Brock. So it was enough for us to go back to those previous studies and submit the relevant data to statistical analysis.

The Results, or Transportation Is Not All

The studies from which we could, and did, extract the transportation data were the Krajewski Facebook experiment, the Fallaci study, the Walker study, and the experiment on attitudes toward dogs and horses. As you can see, then, we decided to include in our analyses also an experiment where no attitudinal impact was shown (the Walker study, that is). We admit that prima facie this might seem strange. What is the use of analyzing the effects of transportation in a case where there was apparently no attitudinal influence for transportation to mediate? However, there was quite a good and simple reason to do so. Namely, it could turn out that even though this text exerted no influence on attitudes of all the participants who read it, it might have still exerted influence on those of them who were highly transported. Such an effect was observed, for instance, in a widely cited study on fiction and empathy where it turned out that fiction had a positive impact on empathic attitudes only in the case of those subjects who were absorbed by, or transported into, the text (Bal and Veltkamp 2013).

No such effect, however, was to be found in our data. "Am I Blue?" had no influence on attitudes toward the welfare of animals in general even after we took the possible mediation by transportation into account. In other words, it did not work even for those who were more transported into the text. And as for those texts which we had shown to have attitudinal influence, the results were mixed. That is, in the Krajewski study and the Fallaci study, the attitudinal influence was indeed mediated (completely or partially) by transportation. These texts did their work by absorbing their readers: the attitudinal effect would not have been possible, or would have been greatly diminished, had it not been for the transportation of our subjects into the texts. However, no such mediation was found in the attitudes toward horses and dogs study. Apparently, then, the mechanism of the narrative impact we observed must have been different in the case of these two different types of attitudes, which is quite puzzling.

What makes our data even more puzzling is that it is at odds with some other existing data. For example, note that in the case of the attitudes toward dogs and horses study the thematic relation between the experimental text and the attitudinal items in the questionnaire was closer, or more concrete than in the case of the other studies. That is, those texts were about dogs and horses as were the items, whereas in the other studies, the texts were about specific animals, while the items concerned animals in general. This is an important detail given that in their 2000 study Green and Brock showed that when the thematic relation between texts and beliefs influenced by those texts was less close, there were fewer cases of that influence being mediated by transportation (707).

Now if you asked us how to explain the divergence that our results point to, we would have to reply that we do not have a strong explanation ready. But even though we cannot explain it, it still allows us to conclude in no ambiguous terms that at least in the case of animal narratives being lost in a story is not always essential to finding it persuasive.

The Crueler the Story, the Better It Makes You

The Discreet Charm of Suffering

Suppose we asked you to name the most engaging stories you know; and these could be any kind of stories – fictional, non-fictional, literary, cinematic, theatrical, and so on. While there are many things that we would not be able to predict about your examples, we can be certain of at least one thing: some, if not most, of them would involve depictions of considerable suffering. Something like, say, despair, trauma, injury, illness, or even death. We are also pretty certain that this would be the case if we asked you about the stories you have enjoyed the most. Your best movies, best novels, and the like. Again, our bet is that in many cases, the plot would involve suffering too.

Now, we are aware of the fact that you might be reluctant to actually admit that these are some of your favorite stories or ones you have enjoyed the most. After all, there is something odd about the idea of enjoying somebody else's suffering, even if it is merely a represented suffering, and even if the representation is fictional. But if you are worried that this makes *you* odd, then we hasten to add that you are not alone. It seems that most people find aesthetic enjoyment in stories that involve suffering, and it might be even the case that these are the stories people enjoy most of all.

Consider what you can find in the so-called greatest literary stories of all time, those that people still read for pleasure, that they constantly return to and cannot live without. Shakespeare? Hate, envy, betrayal, murder, and suicide. Dickens? Poverty, starvation, sickness, exploitation, and death (Carey 1973; Foakes 2003). And we do not even have to remind the readers what atrocities they can find in Dostoyevsky, Tolstoy, Zola, Poe, Kafka, Steinbeck, Nabokov or Orwell. Interestingly, the element of suffering is fundamental even in those of the world's most popular stories that are supposedly less serious than these classics, even in those literary stories that are considered to be the stuff of pure entertainment.

Consider the *Harry Potter* series, which, however strange this may seem at first look, is a narrative brimming with suffering of all kinds. Let us begin with the fact that here we have an orphaned boy who spends the first 11 years of his life in a family of relatives who openly hate, despise, and abuse him on a daily basis. This would of course be a terrible life for any boy to live, but it is further exacerbated in Harry's case by the fact that he possesses some innate magical powers of whose nature he is unaware, but which manifest themselves spontaneously from time to time to his bafflement and alienation. A nightmare indeed (Rowling 2015).

Of course, at some point this changes as he finally comes to accept that magical part of himself having learned that he is a wizard, and that there are more people like him. But then he only gets into worse troubles. First, he also learns that his parents were brutally murdered by the most evil and powerful wizard in the world, that his mother sacrificed her life to defend him, and that the thing that this wizard happens to crave the most is to murder Harry himself. Aside from this psychological burden and the resulting distress, Harry has then to suffer psychologically and physically as a result of the evil wizard's attempts on his life and the hostility of those who serve the latter, including all sorts of vile monsters. He is constantly in fear for his life, he receives serious wounds, his limbs get broken, and more than once his condition is deadly serious.

As if that were not enough, the attacks also threaten the life and well-being of people whom he loves, which results in his nagging feelings of guilt. His loved ones too end up getting seriously wounded and find their lives in danger. Indeed, some of these people actually lose their lives, as do many of the villains, as well a number of innocent bystanders. Blood flows, limbs are severed, and bodies are torn to pieces or burnt. There is despair, panic, hopelessness, betrayal, and then there is prejudice and exploitation. All this in books that children and adults love so much that the series sold 450 million copies having become a very likely contender to the title of the biggest selling book series of all time (Staff 2017).

But wait, not only do children love to read such gruesome stories, they actually like to tell them themselves! In his fascinating book *The Storytelling Animal: How Stories Make Us Human*, Jonathan Gottschall gives the example of a study where preschoolers were asked on the spot to tell a story, and what they produced very often involved things such as

> trains running over puppies and kittens; a naughty girl being sent to jail; a baby bunny playing with fire and burning down his house; a little boy slaughtering his whole family with a bow and arrows; a different boy knocking out people's eyes with a cannon; a hunter shooting and eating three babies; children killing a witch by driving 189 knives into her belly.

Gottschall sees this example as

> support[ing] the play scholar Brian Sutton-Smith, who writes 'The typical actions in orally told stories by young children include being lost, being stolen, being bitten, dying, being stepped on, being angry, calling the police, running away or falling down. In their stories they portray a world of great flux, anarchy and disaster'.
>
> (2012, 34)

The point of invoking the above details about children's stories, Harry Potter, and literary classics was of course to further assure our readers that if they derive aesthetic pleasure from reading about various kinds of suffering, they need not worry that this makes them odd. This *is* the norm. And moreover, it is a norm that is quite sensible from an evolutionary point of view. For a social species such as ours, the most important things are interpersonal interactions and ecological challenges. In our past (but also nowadays), we had to face such problems very often and our ancestors' chances to survive strongly depended on their reactions to them.

But of course, this does not mean that there are no reasons to worry whatsoever. Merely because something is the norm does not mean that it is right. And indeed there have been numerous writers, moralists, and scholars who argue that the aestheticization of suffering and death is morally wrong; that it is wrong to represent suffering and death in an artistic form, and that it is wrong to enjoy such representations aesthetically, regardless of the effects or intentions behind such acts (Grønstad and Gustafsson 2012).

Could they be right? Consider, for instance, Nilüfer Demir's famous photo of the 3-year-old Syrian refugee Alyan Kurdi's dead body lying calmly on a Turkish beach. If it made the impact that it did, having been featured on covers of dozens of magazines and all over the internet, this is in large part because of its aesthetic qualities. Writing for the website of The Ethical Journalism Network, Misja Pekel and Maud van de Reijt observed that "less than a week before [the publication of the photo], the inboxes of photo editors worldwide were bombarded with pictures of seven young drowned children on the Libyan coast. Most newspapers did not publish them." Why? The answer, Pekel and der Reijt argue, "has to do with aesthetics":

> The pictures of the Libyan children are horrific. Their clothes have shifted. Their bodies are evidently lifeless. There is no doubt about the state these victims are in. Apparently, to show horrific events, we need a touch of beauty. Ironic?...
>
> The photo editor of the Dutch newspaper *Trouw* put it this way: "Before, we only saw pictures of decayed bodies. These you simply do not show. Aylan's photo was the first one that made you wonder:

is he asleep or is he dead? That is why we thought it was reasonable to print this picture."

(2017)

But if this is so, then isn't there something sickening in approaching the death of a little boy as a sort of aesthetic spectacle? And doesn't it remain sickening even when we consider that the picture apparently helped to change the European public's minds about immigrants?

For many, the problem persists even if the medium of representation does not involve directly recording the suffering of an actual victim, the way journalistic photography or film do, and even if it does not represent actual people. For instance, *Schindler's List* is not a documentary film, but it still has been objected that its aestheticized representations of the Holocaust are inappropriate (Picart and Frank 2006, 68–69; Kaplan and Wang 2008, 120–24). Similarly, some argued that it was inappropriate both for the makers of *Slumdog Millionaire* to aestheticize poor Indian children in their movie, and for its audience to derive aesthetic pleasure from a story whose salient element was that poverty. It is precisely for this reason that *Slumdog Millionare* has been called "poverty porn" (Lim and Garrett 2009).

While such objections are most often raised with regard to human suffering, they have been also aimed at depictions of the plight of animals (Aaltola 2014; Taylor 2016), and in conducting our research, we had always been aware that they apply to our texts as well. Recall how happy we are about the fact that the narrative which Krajewski wrote according to our suggestions then became part of his novel, that the novel became a bestseller, and that therefore the narrative was read by thousands. But one consequence of this was that it must have been very often read for pure entertainment. Just imagine the very likely case of somebody who reads *The Lord of the Numbers* at a beach reclined on a sun lounger enjoying the gentle rays of the sun and sipping margaritas. For such a person, our animal story is supposed to be, and most likely is, something that complements their perfectly pleasurable day. But then again, this is a story of an animal that is forcibly separated from her mother, burned with cigarettes, and then cruelly tortured by a psychopath!

We said "most likely" because for some readers the animal story might have ruined their mood precisely because of the suffering it involved. Some of them might have even felt brutalized by it. This raises an ethical problem of a different kind. Was it ethical on our part to expose our subjects to potentially distressing material? Either way, in choosing narratives of animal suffering for our material we had to face some serious ethical objections. Yet if we clung on to our stories of suffering despite all of this, it was not because we were evil or sadistic but for the simple reason that we took the available historical and psychological data to suggest that they were most likely to impact attitudes.

Perhaps, however, we were wrong here. Maybe we could have used for our purposes stories that did not involve suffering, let alone the extreme horror of the kind depicted by Eisnitz or Krajewski? After all, there have been studies which show that positive stories can induce story-consistent beliefs too (Green and Brock 2000), and we could imagine stories that might do so in our case. For instance, couldn't a story which portrayed animals as sensitive creatures raise our concern for their welfare? Or a story that stressed how similar we are to animals in this regard? Or one which presented a positive, harmonious relationship between animals and humans?

Somewhat bothered by the aforementioned ethical objections, we felt that we at least needed to test experimentally if this was possible. That is, we decided to conduct an experiment that would allow us to see whether a story that met the above criteria (i.e. one that did not involve violence, but rather a positive relationship between humans and animals, and one that would stress similarities between our mental life and theirs) could impact attitudes toward animal welfare the way our previous stories did, and to compare that influence with that exerted by one of those more distressing tales.

The Experiment and Results

Our experiment was conducted on a group of 108 subjects (67 women) aged 18–19. They were randomly assigned to three groups, including a control group.[4] Consistent with the purpose of the study, one of the experimental groups read a text that we thought would be a good example of a narrative depicting a positive relationship between an animal and a human being. The narrative additionally involved topics that are relevant to the question of moral welfare, such as the capacity of humans to empathize with other animals.

The story came from a popular book *Considering the Horse* by Mark Rashid (2014), who is, as his website states, an "internationally-acclaimed horse trainer known for his ability to assess situations from a horse's point of view" ("Mark Rashid" 2017). He also has some literary talent, as exemplified by his widely popular nine books. We actually were given a hint about that particular book by a horse lover who knew the book in its Polish edition, and the hundred-plus enthusiastic reviews on Amazon only strengthened our opinion that this was the kind of text we had been looking for.

The tale we picked from that book presented the teenage Rashid's first encounter with a particular horse. The two are initially unsure how the other party will react, and therefore anxious, but eventually through a series of small gestures, which Rashid's narrative portrays in detail, they develop a kind of mutual understanding, and the episode ends with apparent relief and satisfaction on the part of both the horse and the man. This is what the first experimental group read about.

The other experimental group read our excerpt from Eistnitz's *Slaughterhouse*. We chose this text because it matched Rashid's in terms of the species depicted therein, but also because, of all our experimental stories that concerned horses, this was the one which had the highest attitudinal impact. But given that during the time which had passed since the last experiment in which we had used it we grew worried about that text's depictions of brutality, we decided to manipulate it slightly. The modification consisted in extracting some drastic details altogether from the text and rewording the text using euphemisms. And thus, for instance, the practices of sticking the knife into a horse's rectum and skinning a horse's head alive were entirely omitted, while a line such as "Run and cut his throat" would be rewritten as "Run and finish him with a knife." We thought that the resulting narrative was less violent than the original one, but of course we could not rely solely on our judgment. We therefore asked several other people, whom we believed to be acquainted fairly well with literary texts, about their opinion, and they confirmed the modification worked as we planned. In psychology, this method is usually referred to as consulting "competent judges."

Assured by our judges we then conducted our experiment. Subjects in each of the groups read their text and were then asked to complete a questionnaire that measured their attitudes and impressions from the text. We hypothesized that Rashid's story might indeed influence attitudes, but that its influence would be less significant and strong than that exerted by the *Slaughterhouse* story.

This hypothesis was based on a 1981 study by Shelton and Rogers which established that footage depicting cruel whaling practices had a greater impact on attitudes toward helping whales than footage showing those animals in their natural environment. The researchers behind this experiment explained this result by referring to their proposed extension of so-called "fear motivation theory," which states that "when a danger threatens us personally, our motivation to protect ourselves" is to a significant extent the "function of … our cognitive appraisal of … the severity of the danger." According to Shelton and Rogers, their results show that this mechanism generalizes to the danger posed to others, so that "the more injurious the potential consequences, the more we are motivated to protect whomever, or whatever, is threatened" (Shelton and Rogers 1981). This sounded reasonable to us, also from an evolutionary point of view,[5] and therefore we expected the impact of our two narratives to differ accordingly.

However, we were in for a surprise. The data obtained in the experiment showed that, contrary to what we had presumed, the *Slaughterhouse* story did not have a stronger attitudinal impact than Rashid's tale. More than that, the data showed also that it did not have any attitudinal influence whatsoever. The attitudes of the subjects who read that version of the text were not better in a statistically significant way than those of the people from the control group. The same applied to Rashid's tale of developing interspecies empathy. We found it had zero influence. Since Eisnitz's text had been previously shown to influence attitudes, we replicated our suffering/non-suffering study. The results were the same.

No impact on attitudes exerted by the softer version of Eisnitz's narrative, and no impact exerted by Rashid's story.

Obviously, the data about Eisnitz was more troubling to us than that pertaining to Rashid. We had to somehow explain the difference between the results obtained in our first Eisnitz study (described in Chapter 3) and these two we had conducted. Fortunately one likely explanation readily suggested itself. After all, in our current experiments, we used a *different* version of the excerpt from the *Slaughterhouse*, one that was less violent than the original one. Could it be, then, that the more suffering a story involved the greater were its attitudinal impact? At least this is what seemed to be suggested by our results, and since it appeared we had just stumbled upon an important mechanism of the narrative influence we were interested in, we decided to add to our schedule an experiment that could confirm that.

The Good, the Bad, and the Cruel

As we agreed above, stories that people tend to enjoy the most very often involve suffering or its threat. What we should now add to this is that the threat featured in these stories is very often extreme to a hyperbolic degree. For instance, note that the evil wizard who is after Harry Potter is not just *any*, slightly malign wizard. He is *the* most evil of all evil wizards and *the* most powerful of them too. So evil and so powerful, indeed, that most characters in the novel do not even dare to mention his name, referring to him as "He-Who-Must-Not-Be-Named." In other words, there is no force in the magical world that could be more dangerous to Harry, or to anyone else for that matter, yet this is precisely the force that is opposed to our protagonist. This pattern is repeated in many other ways throughout the novels. For instance, when Harry has to face one of a group of three dragons, the one he eventually faces is the most dangerous of them. When he is to face the evil dementors, he has to face a whole horde of dementors, and in general the phenomena he is threatened by rank among the most dangerous in their respective categories.

Of course, this is not only the case with Harry. Protagonists of popular books and movies usually face villains that are the most villainous of all, and the danger these villains pose is always ultimate. They either want to destroy the world or to enslave it, and they have the means to do it. Similarly to our Harry Potter examples, that tendency toward hyperbole is again visible on all levels of the plot. For instance, if a character is in danger of falling off a cliff or the top of a building, this always has to be an extremely high cliff or a skyscraper, and if somebody wants to kill our protagonist, the death is usually extraordinarily horrible, involving things like dismemberment, decapitation, or – to quote from the movie *The Hobbit* – "laceration, evisceration, incineration" (Jackson et al. 2014). Recall also that when the owner of Beautiful Joe wanted to punish the dog, not only did he want to hurt it physically, but hurt it permanently by cutting its ears and tail, and not only did he want to cut its tail and ears, but also do it with an axe and with no anesthesia. Extreme danger, always, everywhere, and for everyone, people and animals alike.

One might complain about this, but if one of the goals of such stories is to create suspense, to keep their reader, viewer, or listener on the edge of their seat, then such measures are quite understandable. The bigger the danger the protagonist is in, the more you will be anxious about his or her fate, the same way the extent to which you are worried about a threat to your welfare depends on the severity of the danger in question. Could such mechanisms also influence the extent to which stories impact our attitudes, including attitudes toward animals? It would seem so from Shelton and Rogers's extension of the fear motivation theory. After all, as they argued, the attitudinal impact of a given representation of danger will depend on "the magnitude of noxiousness of [that] danger" (1981, 367).

One good way to see if this was true in the case of our narratives was to compare experimentally the influence of the two versions of Eistnitz's narrative that we had at our disposal. We did precisely this, adding to our design yet another version of the text, one which was still less violent than our previous softened version. Those of the disturbing details from the original version that were still remaining in that moderately violent version were either rewritten in an euphemistic way or simply left out where rewriting was not an option. The resulting variant was as little violent and disturbing as we thought possible.

Our subjects were 89 students of psychology and education (83 of them women), aged 19–27, divided randomly into three groups, each reading a different variant of the text. They were instructed by our confederates about the ostensible subject of the study (which was again to investigate the relations between the personality and worldviews of readers and their reading experiences), and once they have finished reading their assigned story, they were asked to complete a questionnaire. It was similar to those used in studies described in Chapters 2–4, comprising a part which ostensibly measured personality and worldviews and which included the ATAW scale, along with our usual battery of items about impressions from the text and demographical data.

Our results showed that the level of cruelty of the text (or the degree of the noxiousness of the treatment to which horses in the story were submitted) significantly differentiated the influence of the text on attitudes toward animals ($p = 0.02$, $\eta^2 = 0.08$). Consistent with our hypothesis, and the extended fear motivation theory, the story which had the greatest positive attitudinal impact was the original, that is, the one that depicted cruelty most explicitly.

Aside from shedding light on the mechanisms behind the narrative impact we observed previously, this result also allowed us to provide a clear and empirically informed answer to the potential objection that our use of cruel narratives is unethical because so are such narratives and their enjoyment.

And the answer is simply this: while we do realize this approach is not morally unchallengeable, we have good reasons to think that it is effective, and we do think that the positive moral outcomes it leads to may outweigh whatever moral deficiencies it suffers from. Our narratives would then be like a controversial measure that a detective must use to solve an important case. Or like a bitter pill that one must swallow in order to feel better.

But independently of whether you think that developing a more positive attitude toward animal welfare as a result of reading a text counts as a positive moral outcome or not, you might still want to ask what this means in practice, outside of the laboratory and beyond the values of p and η^2 that you see on the page. We are going to answer this question in the next chapter.

Notes

1 Most definitely, too, you could imagine a hypothetical successful detective novel about the chase and catch of Walter White that would not take these factors into account at all.
2 A crime in the USA, Poland, and many other countries in the world (McKenna 2013, 167–70).
3 Here are two other similar cases reported in the media: McNeill (2017) and Robinson (2014).
4 Similarly to our previous studies, the control group read the text about the boson.
5 From an evolutionary point of view, which assumes that the basic mechanisms of our psychology were formed in our evolutionary past (i.e. not only in the last 200,000 years of *Homo sapiens* existence, but also much earlier in our ancestors), fear or danger have to have a stronger effect on us than positive emotions as those former emotions were more important in that period (Foley 1995). Moreover, even apart from that hypothesis, there are reasons to think that natural selection would make negative emotions particularly strong in humans. For instance, in his paper "The smoke detector principle. Natural selection and the regulation of defensive responses," Randolph Nesse makes the following point:

> Defenses, such as flight, cough, stress, and anxiety, should theoretically be expressed to a degree that is near the optimum needed to protect against a given threat. Many defenses seem, however, to be expressed too readily or too intensely. Furthermore, there are remarkably few untoward effects from using drugs to dampen defensive responses. A signal detection analysis of defense regulation can help to resolve this apparent paradox. When the cost of expressing an all-or-none defense is low compared to the potential harm it protects against, the optimal system will express many false alarms. Defenses with graded responses are expressed to the optimal degree when the marginal cost equals the marginal benefit, a point that may vary considerably from the intuitive optimum. Models based on these principles show that the overresponsiveness of many defenses is only apparent, but they also suggest that, in specific instances, defenses can often be dampened without compromising fitness. The smoke detector principle is an essential foundation for making decisions about when drugs can be used safely to relieve suffering and block defenses.
>
> (Nesse 2001)

Works Cited

Aaltola, Elisa. 2014. "Animal Suffering: Representations and the Act of Looking." *Anthrozoös* 27 (1): 19–31. doi:10.2752/175303714X13837396326297.

Bal, P. Matthijs, and Martijn Veltkamp. 2013. "How Does Fiction Reading Influence Empathy? An Experimental Investigation on the Role of Emotional Transportation." *PLOS ONE* 8 (1): e55341. doi:10.1371/journal.pone.0055341.

Carey, John. 1973. *The Violent Effigy: A Study of Dickens' Imagination.* London: Faber and Faber.

Di Bella, Maria Pia, and James Elkins, eds. 2013. *Representations of Pain in Art and Visual Culture.* Routledge Advances in Art and Visual Studies 4. New York: Routledge.

"Driver Reading Behind the Wheel Causes Crash." 2017. *Car Crash Blog*, Accessed July 12, 2017. www.pagelaw.com/car-crash-blog/distracted-driving-accidents/washington-county-crash-caused-by-driver-distracted-by-a-book/.

"Eagan Police Ticket Man For DWR: Driving While Reading." 2017. Accessed July 12, 2017. http://minnesota.cbslocal.com/2016/01/21/eagan-police-fine-man-for-dwr-driving-while-reading/.

Foakes, R. A. 2003. *Shakespeare and Violence.* Cambridge; New York: Cambridge University Press.

Foley, Robert. 1995. "The Adaptive Legacy of Human Evolution: A Search for the Environment of Evolutionary Adaptedness." *Evolutionary Anthropology: Issues, News, and Reviews* 4 (6): 194–203. doi:10.1002/evan.1360040603.

Gottschall, Jonathan. 2012. *The Storytelling Animal: How Stories Make Us Human.* Boston, MA: Houghton Mifflin Harcourt.

Green, Melanie C., and Timothy C. Brock. 2000. "The Role of Transportation in the Persuasiveness of Public Narratives." *Journal of Personality and Social Psychology* 79 (5): 701–21.

Grønstad, Asbjørn, and Henrik Gustafsson, eds. 2012. *Ethics and Images of Pain.* Routledge Advances in Art and Visual Studies 1. New York: Routledge.

Herbert, Rosemary. 2003. *Whodunit? A Who's Who in Crime & Mystery Writing.* New York: Oxford University Press. http://site.ebrary.com/id/10266427.

Jackson, Peter, Ian McKellen, Martin Freeman, Richard Armitage, and J. R. R. Tolkien. 2014. *The Hobbit: An Unexpected Journey.*

Johnson, Dan R. 2016. "Transportation into Literary Fiction Reduces Prejudice against and Increases Empathy for Arab-Muslims." *Scientific Study of Literature* 3 (1): 77–92. doi:10.1075/ssol.3.1.08joh.

Kaplan, E. Ann, and Ban Wang, eds. 2008. *Trauma and Cinema: Cross-Cultural Explorations.* Pbk. ed. Hong Kong: Hong Kong University Press.

Lim, Dennis, and Brandon L. Garrett. 2009. "What, Exactly, Is Slumdog Millionaire?" *Slate*, Accessed January 26, 2009. www.slate.com/articles/arts/the_oscars/2009/01/what_exactly_is_slumdog_millionaire.html.

Macdonald, James S. P., and Nilli Lavie. 2011. "Visual Perceptual Load Induces Inattentional Deafness." *Attention, Perception & Psychophysics* 73 (6): 1780–89. doi:10.3758/s13414-011-0144-4.

"Mark Rashid." 2017. Accessed May 19, 2017. www.markrashid.com/about/about-mark.

McKenna, Erin. 2013. *Pets, People, and Pragmatism.* First edition. American Philosophy. New York: Fordham University Press.

McNeill, Heather. 2017. "Woman Caught Reading a Book While Driving 'down Tonkin Highway.'" *WAtoday*, Accessed April 7, 2017. www.watoday.com. au/national/western-australia/woman-caught-reading-a-book-while-driving-down-tonkin-highway-20170407-gvg0n6.html.

Nell, V. 1988. *Lost in a Book: The Psychology of Reading for Pleasure.* New Haven, CT: Yale University Press.

Nesse, R. M. 2001. "The Smoke Detector Principle. Natural Selection and the Regulation of Defensive Responses." *Annals of the New York Academy of Sciences* 935 (May): 75–85.

Pekel, Misja, and van den Reijt Maud. 2017. "Refugee Images – Ethics in the Picture." Ethical *Journalism Network*, Accessed July 12, 2017. http://ethicaljournalismnetwork.org/resources/publications/ethics-in-the-news/refugee-images.

Picart, Caroline Joan, and David A. Frank. 2006. *Frames of Evil: The Holocaust as Horror in American Film.* Carbondale: SIU Press.

Plous, Scott. 2003. "Is There Such a Thing as Prejudice toward Animals?" In *Understanding Prejudice and Discrimination*, edited by Scott Plous, 509–28. Boston, MA: McGraw-Hill.

Rashid, Mark, and Ron Ball. 2014. *Considering the Horse: Tales of Problems Solved and Lessons Learned.* New York: Skyhorse Publishing Inc.

Robinson, By Julian. 2014. "Woman Who Was Photographed Driving Ten Miles on Busy Motorway Reading BOOK at the Wheel Smiled and Waved When Another Furious Driver Beeped at Her." *Mail Online*, Accessed May 30, 2014. www.dailymail.co.uk/news/article-2643934/Woman-photographed-driving-ten-miles-busy-motorway-reading-BOOK-wheel-smiled-waved-furious-driver-beeped-her.html.

Rowling, J. K. 2015. *Harry Potter and the Sorcerer's Stone.* First illustrated edition. New York: Arthur A. Levine Books, an imprint of Scholastic Inc.

Rzepka, Charles J., and Lee Horsley, eds. 2010. *A Companion to Crime Fiction.* Blackwell Companions to Literature and Culture 66. Chichester; Malden, MA: Wiley-Blackwell.

Sauerberg, Lars Ole. 2016. *The Legal Thriller from Gardner to Grisham: See You in Court!* Crime Files. London: Palgrave Macmillan.

Shelton, Mary Lou, and Ronald W. Rogers. 1981. "Fear-Arousing and Empathy-Arousing Appeals to Help: The Pathos of Persuasion." *Journal of Applied Social Psychology* 11 (4): 366–78. doi:10.1111/j.1559-1816.1981.tb00829.x.

Staff, TIME. 2017. "Because It's His Birthday: Harry Potter, By the Numbers." *Time*, Accessed July 12, 2017. http://entertainment.time.com/2013/07/31/because-its-his-birthday-harry-potter-by-the-numbers/.

Taylor, Nik. 2016. "Suffering Is Not Enough: Media Depictions of Violence to Other Animals and Social Change." In *Critical Animal and Media Studies: Communication for Nonhuman Animal Advocacy*, edited by Núria Almiron, Matthew Cole, and Carrie P. Freeman, 42–55. Routledge Research in Cultural and Media Studies 77. New York: Routledge, Taylor & Francis Group.

7 How Long Will It Work?

A Short Chapter on Attitudinal Impact Over Time

How to See Both the Forest and the Trees

We began this book by extolling the virtues of narrowing down the subject of one's investigation, and while we still stand by that general assessment, we would like to add a bit of nuance by saying something about the vices of doing so. For instance, if you are a detective trying to fight a large criminal network, it would be futile to do so by trying to investigate the whole network as such. A network like that would simply be too big and nebulous for one detective or even one investigative team to grapple with, let alone to expose and compromise.

A better way would be to start by investigating its smaller sub-network, or even a single member. Then, you would be dealing with something clearly identifiable, trackable, and that can be apprehended. Of course the problem then is that the more you are absorbed by that smaller, more manageable task, the more you lose the sight of the bigger phenomenon that prompted your investigation in the first place. So you might then feel that having achieved that task, apprehending that individual, you can rest fully satisfied, while this would in fact be only the beginning of a larger job. The suspect might be caught, but the network would still be operating.

Ditto for our investigation. What inspired us was the larger question of whether narratives can contribute to reducing animal suffering in our societies by changing the public's minds. Recall we had agreed that framed this way, the question was hard to study. It was simply too broad and its contours problematically indistinct. Recall too, that we had therefore chosen to focus on its smaller element, the influence of stories on attitudes toward animal welfare. As we hope the preceding pages testify, chasing thus defined suspect demanded a lot of effort, ingenuity and time, and we also hope the readers will share with us the feeling that such a chase could be absorbing. But, in being absorbed by it and in enjoying the progress of our investigation, we might have easily forgotten how far we still were from the larger and more important issue with which we had begun. We might have failed to see the forest for the trees. Here is why.

Note that all of those experiments described in the preceding pages, which consumed a great deal of time and effort, focused merely on

DOI: 10.4324/9780429061424-8

measuring attitudes *immediately* after reading a story. While such results can be justifiably said to describe attitudinal change, there is one important limitation to them. Namely, the existing psychological literature does not allow us to simply *presume* that the attitudinal effect will last any longer.

What we mean by the existing psychological literature here is mainly the research on so-called priming. As we already said in Chapter 4, in priming, the exposure to a given factor stimulates certain associations or memories in one's mind that then affect the way we react to another stimulus. This phenomenon had been initially discovered in studies on linguistic comprehension where it turned out that subjects read a given word (e.g. "nurse") faster if shortly before that they read a word with similar connotations (e.g. "doctor"). The explanation was that the latter word somehow prepared the subjects to read the former word by activating the respective parts of their memory. In other words, they were primed to read certain words faster. From then on priming has been shown to exist in many other spheres of life and to have fantastically interesting consequences, which led to the enormous impact of that notion in psychology (Molden 2014).

While that influence is no longer as significant as it used to be, thanks to the fact that the results of some studies on priming could not be replicated while some others turned to be fraudulent,[1] there is still a place for priming in psychological research, and some of that research has important implications for our project. Namely, our subjects might be seen as primed by animal stories in such a way that the emotional state in which these stories put them made them answer the animal-related questions differently than they would have had they not been in that state. But how long do you remain in an emotional state in which a story puts you? Skeptics could argue that state may well be gone an hour later, or even after 15 min. If the attitudinal change induced by the narrative depended on that state, it would naturally be gone within that time span, too. And that, for us, would be a very disappointing outcome.

But on the other hand, judging by our own experiences, there are stories which hold us in their emotional grip for a long time. Stories we cannot get out of our heads, stories which haunt us on an on, and most of these stories happen to concern suffering. Perhaps, then, our animal stories, which commonly concern the plight of animals, could exert their attitudinal influence over a longer time *even if* that influence depended on an emotional state they put us in? Or perhaps, the attitudinal effect of our stories depended on a still different mechanism, one that would not be at all subject to the limitations described by the theory of priming?

In any event, we could not just leave these questions hanging like that. We had to be sure. If what we are concerned with is whether stories can induce an actual social change with regard to the ill-treatment of non-human animals, we simply have to know whether their impact is lasting. After all, the main reason why people still talk about *Black Beauty* is that its readers did join humane societies and did push their representatives to do such things, and that kind of influence lasted for decades. In order to address these limitations, we conducted two additional experiments, one measuring the impact of stories after a week, and the other measuring this impact over a period of up to two months.

The Experiment and Results: After One Week

Our participants in this study were 62 high school students (29 women), aged 18–19, and the study itself took place in improvised laboratory spaces at the high school they attended. The subjects were randomly divided into two groups and primed about the ostensible purpose of the study, which was, as usual, the relation between the readers' worldviews and personality and how they experience texts. They were informed that they were about to read a text and that we would ask them to fill out a special questionnaire a week later. Subjects in the experimental group then proceeded to read Krajewski's story about Clotho that we used in our previous studies, while the control group read the story about the Higgs boson particle. A week later, the subjects were invited to the same spaces in which they read the text and asked to fill out the questionnaire. The questionnaire comprised 72 items and was similar to that used in our Facebook experiment and the experiments described in Chapters 3 and 4. Once they had filled it in, the participants were then thanked for their participation and dismissed.

What our results revealed was that there was a significant even if weak main effect of the experimental condition ($\eta^2 = 0.07$). The participants from the experimental group expressed more pro-animal welfare attitudes than participants from the control group. We have established, then, that the positive impact of a fictional narrative on attitudes toward animal welfare can be observed as long as a week after exposure.

In doing so, we corroborated the results obtained in our previous studies: what our present data implied was that the previously observed effects were definitely not fleeting. But they also implied something more. After we obtained them, we returned to the details of the Facebook study, in which we measured the impact of the same text immediately after reading, and compared the strengths of the effects. We were curious if there would be any difference, even if comparing the results of experiments as different as those two has its limitations. In any event, it turned out that the strength of the effect of that same text after a week ($\eta^2 = 0.07$) was greater than that observed immediately after reading ($\eta^2 = 0.02$). But why? Shouldn't it actually be smaller?

This made us think about the so-called absolute sleeper effect hypothesis as applied to fiction (Appel and Richter 2007). This variant of the absolute sleeper hypothesis is based on the following three premises. (1) The reader's awareness that a text is fictional acts as a discounting cue and mitigates the persuasive influence of the text's content. (2) The reader's memory of the text's fictionality "decays relatively fast or is dissociated from the memory of the content of the narrative" (Appel and Richter 2007, 118) (3) This decay or dissociation tends to occur while the memory of the content remains stable (cf. Radvansky and Zacks 2014).

What follows from these premises is that at some point after reading the persuasive influence of the fictional story should actually increase. This should occur precisely at the point when the discounting cue is already gone but the memory of the content remains unchanged. In some cases, the mitigating effects of the awareness of fictionality may even be so strong that the persuasive influence of a fictional text will only be observable after some time, and not immediately (Appel and Richter 2007, 118). This hypothesis has been confirmed experimentally, including in a study which showed that participants' empathy increased a week after exposure to a fictional text (Bal and Veltkamp 2013).

It seemed, then, that our results corroborate this variant of the sleeper hypothesis, which would be an interesting additional contribution of our research. But unfortunately, the sleeper hypothesis is in tension with the results obtained in our study on fictionality. Recall that, in that study, we showed that there was no difference in attitudinal impact between two identical texts, one of which was clearly presented to the subjects as fictional and the other not. That experiment apparently showed that fictionality does not function as a discounting cue at least as far as attitudes toward animals are concerned.

However, it is also true that the experiment and the present one involved different texts. Perhaps, the text by Eisnitz (which we used in the study on fictionality), while still having features that allowed us to present it believably as fictional, was in itself less characteristic of fictional texts than the story by Krajewski? The latter story was taken from a novel, after all, involved an exotic animal, a mysterious psychopath called the Lord of the Numbers, and a narrator typical of fictional novels. Eisnitz's text involved a first-person narrator, consisted largely of dialogue, and involved no mysterious entities, elaborate plots, and the like. Krajewski's story was definitely more literary in a conventional sense. Perhaps even if we succeeded at priming our readers to believe that a version of the text was fictional, given the features of the text this awareness was not strong enough to act as a discounting cue? Unfortunately, we could only speculate on whether that was indeed the case, and had to leave this mystery as it was. What mattered was our main hypothesis in this study, which was confirmed. The attitudinal impact of animal stories is not momentary – it can last for at least as long as a week.

The Experiment and Results: Up to Two Months

That was a good result, but we wanted to see if we could have a better one: we wanted to see if the influence of that same text could last even longer, that is up to two months after it was read. In hypothesizing that it might last as long as this, we followed one of the few existing studies on the influence of fictional narratives on attitudes toward out-groups over time (Vezzali, Stathi, and Giovannini 2012).[2] In that study, Italian high school students were randomly divided into three groups: the experimental one, whose participants were asked to read their chosen novel on intercultural topics during their summer break (ca. two months), and two control groups, one of whom was asked to read a novel on a topic unrelated to minorities over the same period, and the other who did not read any novel at all. A week after the summer holidays were over, all three groups were asked to complete a questionnaire measuring attitudes toward immigrants. The results indicated that these attitudes had improved in the experimental group relative to the control groups (Vezzali, Stathi, and Giovannini 2012).

Particularly valuable about this experiment was its ecological validity (Kellogg 2002, 120), which as we have already stressed, is often rather low in experimental studies on narratives. Unlike in typical laboratory studies, participants in this one read a book in an environment of their choice, and the way they wanted to. In our study, we attempted to secure the same, or even higher, level of ecological validity, and to this end we again took advantage of our cooperation with Marek Krajewski, in particular of the fact that he agreed to implant our experimental story in his book *The Lord of the Numbers* (2014).

Our subjects in this study were divided into two groups. There was the experimental group, who read *The Lord of the Numbers* within a period of two months immediately after its publication, and the control group, who did not read it within that time span. To assess whether the narrative had an impact on our subjects' attitudes toward animal welfare, we measured those attitudes before the publication of the book and after the end of the two-month period.

The participants were 410 respondents (298 women), aged 18–60, of a custom online panel of a market research agency which we hired to conduct the study. They were selected before the publication of *The Lord of the Numbers* on the basis of their previous experience with Marek Krajewski's work. In order to qualify for the study, they must have read at least one book by that author before the publication of *The Lord of the Numbers*. We used this measure in order to ensure that a sufficient number of the participants in our sample would subsequently read *The Lord of the Numbers* on their own. This was naturally more likely in the case of those people who had read Krajewski's work before than those who had not. The participants were awarded points which they could later exchange for money.

The custom online panel of the agency we cooperated with included individuals from all around Poland, of different ages and backgrounds, who in most cases cannot know one another. Seven weeks before the official date of the publication of *The Lord of the Numbers* (which was released on Sept. 11, 2014), the agency invited its panelists to a study on the worldviews represented by readers of different authors. Eventually, 2,000 people volunteered to participate and they were asked, among other things, about how many books they had read in the previous 12 months, about their favorite kinds of books, which of the 29 listed contemporary Polish authors they knew, and whether they had read any books by those authors.

Those who said they had read at least one book by Krajewski were then asked to fill out a questionnaire, which was similar to that used in the previous study described in this chapter in that it included the part about the worldviews of readers, including the ATAW items, and metrical questions. This way, 410 people took part in the first phase of our study.

The second phase began on Nov 17, 2014, i.e. two months after the publication of *The Lord of the Numbers*, when the subjects were invited to again fill out our questionnaire. This time, however, they were first asked whether they had read any from a list of 17 recently published books, with *The Lord of the Numbers* among them. Given Krajewski's popularity, that question could not have raised suspicions as far as a general survey on the worldviews of readers were concerned.

All the subjects were then asked to fill out the same set of items they answered in the first phase. Those who indicated that they had read the *The Lord of the Numbers* constituted our experimental group. Those who answered that they had not read the novel constituted the control one. We had to exclude the possibility that in the period which had passed since the first measurement the participants were subject to some factor other than the book which might have changed their attitudes toward animal welfare. In this context, 99 people (including 65 women) between 19 and 58 years of age qualified to the experimental group, whereas our control group consisted of 311 people (including 232 women) aged 18–60 years

To verify whether our experimental setting influenced attitudes toward animal welfare, we performed our usual statistical analyses, which showed that the experimental narrative did not have observable results on attitudes toward animal welfare up to two months after exposure. Note here that this result is not inconsistent with the sleeper hypothesis that was corroborated in the experiment with the impact of the story over a week. That hypothesis does not imply that the persuasive effects of fiction will grow in strength indefinitely, but only that they will grow in strength depending on a certain relation between the memory of the text's fictionality and the memory of its content. Moreover, it implies that persuasive effects of fiction will ultimately depend on the memory of the text's content. It is not improbable that for most of our participants in this study, their memory of the book's content was weak enough at the time of the second measurement for the persuasive effects to have dissipated.

Another possible explanation for the lack of effect after two months as compared to the effect we observed after a week was that in the study over the shorter period participants were statistically younger (18–19 years old as compared with the participants of the study over the longer period, who were between 18 and 60 years old), and more importantly, they attended the same school and could in principle discuss the text among themselves in the time which passed between their exposure to the narrative and the measurement. If this actually happened, then this social reinforcement aspect could have made the attitudinal impact of the text stronger than was the case with the participants in the panel study, who did not have a chance to discuss the text with one another.

Whatever the explanation for it, the lack of attitudinal effect after two months may definitely seem disappointing to advocates of using animal stories for pro-animal purposes. But this result certainly does not mean that they should abandon their hopes for the usefulness of narrative persuasion. Narratives *do* exert influence on attitudes toward animals, it is just that the public apparently needs to be exposed to such texts systematically, and periodically in order for the improved attitudes to be present permanently. Finally, let us not forget that we showed in this chapter that the attitudinal impact of a story can last at least as long as a week. What it means in practice, as the available neurological data on brain connectivity in readers suggests (Berns et al. 2013), is that once you have read a story of an animal's plight, this starts a process that seems to run continuously in the back of your mind for at least seven days whatever you are doing at the moment, perhaps whether shopping, dining, watching TV, or sleeping. This, as the readers will agree, is an amazing, perhaps even eerie capacity, and given that among all those things that we do any given week many concern animals in this way or another, we should not underestimate the attitudinal impact that could be had even by a single little story.

Notes

1 For more on this subject, see Lilienfeld and Waldman (2017, 143–44) and Molden (2014, 206).
2 To our best knowledge the only other longitudinal studies on the influence of fiction on attitudes toward out-groups are Vezzali et al. (2015), Alsbrook (1970), cf. Hakemulder (2000). All other available studies on the respective influence of fiction study that influence immediately after reading, see, e.g., Ellithorpe, Ewoldsen, and Porreca (2015), Johnson et al. (2013), Johnson, Huffman, and Jasper (2014), Kaufman and Libby (2012), Mazzocco et al. (2010), Johnson et al. (2013), and Małecki, Pawłowski, and Sorokowski (2016).

Works Cited

Alsbrook, E. Y. 1970. "Changes in Ethnocentrism of a Select Group of College Students as a Function of Bibliotheraphy." PhD Diss., Urbana-Champaign: University of Illinois.

Appel, Markus, and Tobias Richter. 2007. "Persuasive Effects of Fictional Narratives Increase Over Time." *Media Psychology* 10 (1): 113–34. doi:10.1080/15213260701301194.

Bal, P. Matthijs, and Martijn Veltkamp. 2013. "How Does Fiction Reading Influence Empathy? An Experimental Investigation on the Role of Emotional Transportation." *PLOS ONE* 8 (1): e55341. doi:10.1371/journal.pone.0055341.

Berns, Gregory S., Kristina Blaine, Michael J. Prietula, and Brandon E. Pye. 2013. "Short- and Long-Term Effects of a Novel on Connectivity in the Brain." *Brain Connectivity* 3 (6): 590–600. doi:10.1089/brain.2013.0166.

Ellithorpe, Morgan E., David R. Ewoldsen, and Kelsey Porreca. 2015. "Die, Foul Creature! How the Supernatural Genre Affects Attitudes Toward Outgroups Through Strength of Human Identity." *Communication Research*, October, 1–23. doi:10.1177/0093650215609674.

Hakemulder, Jèmeljan. 2000. *The Moral Laboratory: Experiments Examining the Effects of Reading Literature on Social Perception and Moral Self-Concept.* Utrecht Publications in General and Comparative Literature, v. 34. Amsterdam; Philadelphia, PA: J. Benjamins Pub.

Johnson, Dan R., Brandie L. Huffman, and Danny M. Jasper. 2014. "Changing Race Boundary Perception by Reading Narrative Fiction." *Basic and Applied Social Psychology* 36 (1): 83–90. doi:10.1080/01973533.2013.856791.

Johnson, Dan R., Daniel M. Jasper, Sallie Griffin, and Brandie L. Huffman. 2013. "Reading Narrative Fiction Reduces Arab-Muslim Prejudice and Offers a Safe Haven From Intergroup Anxiety." *Social Cognition* 31 (5): 578–98. doi:10.1521/soco.2013.31.5.578.

Kaufman, Geoff F., and Lisa K. Libby. 2012. "Changing Beliefs and Behavior through Experience-Taking." *Journal of Personality and Social Psychology* 103 (1): 1–19. doi:10.1037/a0027525.

Kellogg, Ronald T. 2002. *Cognitive Psychology.* London: SAGE Publications.

Krajewski, Marek. 2014. *Władca liczb.* Kraków: Znak.

Lilienfeld, Scott O., and Irwin D. Waldman, eds. 2017. *Psychological Science under Scrutiny.* Hoboken, NJ: Wiley.

Małecki, Wojciech, Bogusław Pawłowski, and Piotr Sorokowski. 2016. "Literary Fiction Influences Attitudes toward Animal Welfare." *PLOS ONE* 11 (12): e0168695. doi:10.1371/journal.pone.0168695.

Mazzocco, Philip J., Melanie C. Green, Jo A. Sasota, and Norman W. Jones. 2010. "This Story Is Not for Everyone: Transportability and Narrative Persuasion." *Social Psychological and Personality Science* 1 (4): 361–68. doi:10.1177/1948550610376600.

Molden, Daniel C., ed. 2014. *Understanding Priming Effects in Social Psychology.* New York ; London: The Guilford Press.

Radvansky, Gabriel A, and Jeffrey M Zacks. 2014. *Event Cognition.* Oxford: Oxford University Press.

Vezzali, Loris, Sofia Stathi, and Dino Giovannini. 2012. "Indirect Contact through Book Reading: Improving Adolescents' Attitudes and Behavioral Intentions toward Immigrants." *Psychology in the Schools* 49 (2): 148–62. doi:10.1002/pits.20621.

Vezzali, Loris, Sofia Stathi, Dino Giovannini, Dora Capozza, and Elena Trifiletti. 2015. "The Greatest Magic of Harry Potter: Reducing Prejudice." *Journal of Applied Social Psychology* 45 (2): 105–21. doi:10.1111/jasp.12279.

Conclusions, Speculations, and Prospects

As any reader knows, while every story has an ending, only some have a conclusion, in the sense of providing a clear explanation of the major unknowns the plot contained (Baeten 2005). Some stories not only do not provide such an explanation but leave the reader even more perplexed than he or she had been while reading them. They are enigmas without answers. But there are some genres of stories where this is not allowed. One such kind is the detective story, where providing the answer is obligatory, where it constitutes the major goal of the story, and at the same time the major reward for the reader (Cook 2011). Detective stories are like crosswords in this way, and they have been condemned precisely for this reason, as a primitive form of entertainment, by various intellectuals who revel in ambiguities and darkness (Delamater and Prigozy 1997, 97).

The same holds true for research stories, mutatis mutandis. As we pointed out at the beginning of this book, every research project is a temporal affair and therefore its description must always involve narrative elements. It is only natural, then, that many scholarly articles and books are stories of sorts. They contain expositions of a certain plot, which present the goals a given researcher wants to achieve, and then they depict the path he or she has travelled in order to do so. And again, while all such stories have endings, only some have conclusions. Some scholarly papers and books end without giving any definite answers. While we do not a priori deny the value of such scholarly stories, ours is not like those. As we pointed out, it is a kind of scholarly equivalent of a detective story: it revolves around our investigating a suspect and it promises to reveal whether the suspect is guilty. It is now time for us to provide you with the final report from our investigation. To wrap things up, to state clearly what we managed to achieve, and what all this means. Here it is.

The main goal of our investigation was to establish whether the suspect (narratives of the plight of animals) actually does what we had suspected it of doing, that is, whether it improves attitudes toward animal welfare, or makes our attitudes pro-animal. We are then pleased to report that in the course of our investigation, we obtained evidence that

DOI: 10.4324/9780429061424-9

allows us to unambiguously assess whether the suspicion was warranted or not. The evidence comes both from provocation and interrogation, that is, from a series of natural and laboratory studies. Their results clearly indicate that the suspect *is* guilty: animal narratives do improve attitudes toward animals. The results obtained in this series of experiments also provide us with data which add nuance to our picture of the suspect's activities. Among other things, we established that those narratives of animal suffering that are perceived as fictional influence attitudes toward animals as effectively as those which are perceived as non-fictional, and that this influence may depend on the species of its animal protagonist.

In addition to our main goal, we also established that the attitudinal impact of stories is not fleeting even if it also appears not to be permanent: it can last as long as a week after exposure, although it may be gone after two months. Finally, we also gathered evidence about the mechanisms behind the suspect's actions. In light of this evidence, it is very likely that its impact is not necessarily mediated by transportation into text, and that it is dependent on the cruelty and severity of the animal suffering it depicts.

We believe that our research therefore shows that narratives could be *widely* used to improve attitudes toward animal welfare. And by "widely," we mean that they appear to have this effect on people of varying demographic characteristics. Recall that our data came from more than 3,000 people aged between 14 and 81, both women and men, high school and university students, and those who had already graduated from school or university.

By saying that animal narratives can be widely used for improving attitudes toward animal welfare we also mean that, according to our data, they appear to work in a broad range of contexts and settings. As indicated by our natural experiment, they can be effective in almost any circumstances in which someone might freely choose to read a narrative.

In other words, having concluded our investigation, we may assure all the writers, educators, and activists who want to raise the public's concern for animals, that they may well use narratives of animal plight for that purpose. We guarantee positive results.

This would be a very nice conclusion to our report, but we feel we cannot close it without answering at least two major worries that have been expressed on different occasions about our research. One of them is that the narrative impact we observed is not strong enough, while the other complains that it is too strong indeed. Beginning with the former, recall that the attitudinal effect we observed was not very impressive (with the value of η^2 equaling 0.11 at a maximum) and seems to have faded after merely two months. This, we admit, is definitely a far cry from the available data on what particular stories such as *Black Beauty* and "They Die Piece By Piece" could achieve. Their impact on human minds was apparently immense and lasting. Is there not a conflict between these data and our results?

To see why this does not necessarily have to be the case, allow us to speculate a bit. Consider, first, that *Black Beauty* was an immediate massive hit, and this means, by definition, that it was exceptional in its aesthetic appeal. It was, after all, one of the biggest selling books of the nineteenth century, and indeed of all time (Waldau 2013, 140; Davis 2016, 242n7). It is plausible that the powerful attitudinal effect attributed to it is related precisely to its exceptionally powerful aesthetic appeal. Moreover, since being a massive bestseller involves being read by a large population of people, a book like *Black Beauty* had a chance to reach a relatively large number of people who were sensitive to the plight of animals, or prone to prosocial behavior, or both. Perhaps, the social impact observed by historians was generated not by the general body of *Black Beauty*'s readers, but by its fraction which consisted of readers with exceptional prosocial predispositions. Given how large that body was, the fraction was simply big enough for its impact to be visible.

Note that the above effects can account also for the immense impact of *Beautiful Joe* as well as contemporary stories such as "They Die Piece by Piece." This brings us finally to this conclusion that our results do *not* deny that stories can exert that kind of impact. Rather, they indicate that such a strong impact is not common and most likely depends on a collusion of contextual factors that are hard to predict let alone arrange. In other words, one cannot assume such a profound impact for *any* given story. The good news, though, is that one cannot rule it out either.

But this is good news only for some people. For some others, this is indeed very *bad* news, and so it is bad news that stories can influence our attitudes toward animals even at the value of η^2 as low as 0.02. These people's worries are related to the fact that, given all that we have said about narratives appealing to emotions rather than reason, narrative influence seems to be a form of manipulation. Indeed, doesn't making people read a story in order to make them unwittingly change their mind on a certain issue smack of brainwashing, slipping something into their drink, putting them under hypnosis, and the like?

To put it in different terms, what is worrying about the kind of narrative persuasion we studied is that it lacks an explicit warning that somebody is out to change your mind, not to mention that in narrative persuasion, your mind is being changed in such a way that you may not even be aware that it is happening! This naturally leaves you little opportunity to defend your position or assess the validity of the one you are being induced to take. Given all this, our narrative measures of changing people's moral views on animals may itself be considered immoral. Should that worry us?

If it worries you, then we would like to point out that this is probably because the common notion of manipulation involves such things as making somebody unwittingly develop beliefs or attitudes that are some-how *harmful* to that person and *beneficial* to the manipulator. If these are your criteria for manipulation then we would like to stress up front that our narrative persuasion does not meet them, or at least most of them.

For one, we think that there is no harm for our subjects in developing those attitudes. Second, although we do not have time to argue for this here, we believe that having better attitudes toward animals not only make you a better person but will also be psychologically beneficial for you. Third, we did not seek to persuade our subjects to believe in some-thing that would be directly beneficial for us, the "manipulators." The main beneficiaries of the narrative impact we observed was intended to be animals. To sum up, then, our intentions were clean and so are our hands.

But one might object that despite all that, the kind of narrative per-suasion we advocate is still manipulative in the sense that readers are unaware that somebody is attempting to change their minds in a funda-mental way (Hogan 2009, chap. 4; cf. Oatley 2011, 174;). They are not aware of what is being done to them and therefore unable to react. They are defenseless, and you might indeed say that this is morally wrong. Our reply to this would be that to morally question the practice of us-ing stories to increase concern for the welfare of animals solely on such grounds would entail that using stories for changing attitudes to any other socially important issues is morally dubious too.

Yet, as we already mentioned in the "Introduction," the practice of using literary stories to change such attitudes has been widely accepted, even venerated, in Western culture for thousands of years (Booth 1988; Nussbaum 1990; Keen 2007; Pinker 2011). More than that, the practice has been also shown to lead to undeniably desirable social outcomes in a number of cases. Recall our example of *Uncle Tom's Cabin*, which is widely acknowledged to have contributed to the abolition of slavery through its portrayals of the plight of African Americans (Ammons 2007; Morris, Sachsman, and Rushing 2007). Today, novels such *as The Color Purple* and *Beloved* are widely used to fight racial prejudice, and are even assigned for that purpose as required readings at schools (Hinton and Dickinson 2007). Of course, again, the fact that a given practice is widely used and celebrated does not entail that it cannot be morally wrong. But the above examples strongly suggest that the poten-tial good that might be achieved with the help of narrative persuasion can outweigh whatever is morally questionable with using it for that purpose, especially given that the persuasive potential of rational argu-mentation may simply be lower in some cases than that of stories. But having said that, we have to admit that since we can rely here only on in-direct empirical and theoretical evidence, we cannot tell you if it actually *will* be lower. In order for us to know that, we would have to conduct a separate series of experiments.

This is probably a good occasion to state clearly that separate experiments would also be needed in order to know some other things that are crucial for fully assessing the practical potential of the narrative impact we observed. For instance, we boasted quite a lot about the generalizability of our results, but even those who will appreciate the size of our samples and their demographic diversity, may still point out that they were *culturally* uniform. Wouldn't it be important from a practical point of view to conduct parallel studies in different cultural contexts, e.g., in countries where attitudes toward animals or stories would be different from those in Poland?[1] By the same token, would it not be useful to compare the impact of literary stories with stories told in different media, especially those which surpass literature in popularity such as film and video games?[2]

Finally, ours was a project on attitudes and so is this book. But note that all the large claims about the pro-animal value of stories presume they can affect not only attitudes but also behavior. Apart from making people think differently, *Black Beauty*, "They Die Piece by Piece," and *Babe* apparently made them *do* certain things – join a humane society, send a letter to a newspaper, change their diet. Our research does not say anything about such things, and while the existing psychological literature suggests that they are likely, it also advises caution. For it does point to numerous cases where attitudes do not predict behavior.[3] In other words, we cannot be sure whether the attitudinal impact we observed translates into a behavioral one. Would it not be useful then to study experimentally whether animal stories can also influence actions?[4]

The answer to all of the above questions is of course yes, which brings us to the one final general truth about investigations that we would like to address in this book, which is this, that every investigation always points to new cases that need to be solved. The investigative job of detectives or researchers is never done, and the more they learn, the more they recognize how little they know still. On the one hand, this may be frustrating, but on the other it is reassuring, as it means that there will always be new and exciting perspectives opening before detectives, researchers and anyone whose job consists in investigating. It is also reassuring for those who like to read about scholarly or detective investigations because it means that there will be more stories coming along. We can only hope that the one we told in this book, which is now finally over, was gripping enough for our readers to look forward to more stories from us.

Notes

1 Also, it is worth noting that our studies did not concern children's attitudes toward animals. See, e.g., Eagles and Muffitt (1990), Kellert and Westervelt (1983), and Ascione (1992).

2 Cf. Mossner (2014), Swim and Bloodhart (2015), Ingram (2000), Moore (2016), Schutten (2008), Arendt and Matthes (2016), Jr (2012), Louw (2006), Scott (2003), Bagust (2008), and Kalof et al. (2016).

3 Cf. Glasman and Albarracín (2006), Ajzen (2005), Kraus (1995), Batson et al. (1997), and Heberlein (2012).

4 Cf. Freeman (2010), Cornelisse and Sagasta (2018), Samuels, Meers, and Normando (2016), Braunsberger (2014), and Smith, Ham, and Weiler (2011).

Works Cited

Ajzen, Icek. 2005. *Attitudes, Personality, and Behavior.* 2nd ed. Mapping Social Pyschology. Maidenhead, Berkshire, England; New York: Open University Press.

Ammons, Elizabeth. 2007. *Harriet Beecher Stowe's Uncle Tom's Cabin: A Casebook.* Oxford: Oxford University Press.

Arendt, Florian, and Jörg Matthes. 2016. "Nature Documentaries, Connectedness to Nature, and Pro-Environmental Behavior." *Environmental Communication* 10 (4): 453–72. doi:10.1080/17524032.2014.993415.

Ascione, Frank R. 1992. "Enhancing Children's Attitudes about the Humane Treatment of Animals: Generalization to Human-Directed Empathy." *Anthrozoös* 5 (3): 176–91. doi:10.2752/089279392787011421.

Baeten, Jan. 2005. "Closure." In *Routledge Encyclopedia of Narrative Theory,* edited by David Herman, Manfred Jahn, and Marie-Laure Ryan. London; New York: Routledge.

Bagust, Phil. 2008. "'Screen Natures': Special Effects and Edutainment in 'New' Hybrid Wildlife Documentary." *Continuum* 22 (2): 213–26. doi:10.1080/10304310701861564.

Batson, C. Daniel, Eddie Harmon-Jones, Heidi J. Imhoff, Erin C. Mitchener, Lori L. Bednar, Tricia R. Klein, Lori Highberger, and Marina Polycarpou. 1997. "Empathy and Attitudes: Can Feeling for a Member of a Stigmatized Group Improve Feelings toward the Group?" *Journal of Personality and Social Psychology* 72 (1): 105–18. doi:10.1037/0022-3514.72.1.105.

Booth, Wayne C. 1988. *The Company We Keep: An Ethics of Fiction.* Berkeley: University of California Press.

Braunsberger, Karin. 2014. "The Impact of Animal Welfare Advertising on Opposition to the Canadian Seal Hunt and Willingness to Boycott the Canadian Seafood Industry." *Anthrozoös* 27 (1): 111–25. doi:10.2752/17530 3714X13837396326530.

Cook, Michael. 2011. *Narratives of Enclosure in Detective Fiction: The Locked Room Mystery.* Crime Files Series. Houndmills, Basingstoke; New York: Palgrave Macmillan.

Cornelisse, Tara M., and Jacquelyn Sagasta. 2018. "The Effect of Conservation Knowledge on Attitudes and Stated Behaviors toward Arthropods of Urban and Suburban Elementary School Students." *Anthrozoös* 31 (3): 283–96. doi:10.1080/08927936.2018.1455450.

Davis, Janet M. 2016. *The Gospel of Kindness: Animal Welfare and the Making of Modern America.* Oxford; New York: Oxford University Press.

Delamater, Jerome and Ruth Prigozy, eds. 1997. *Theory and Practice of Classic Detective Fiction.* Contributions to the Study of Popular Culture, no. 62. Westport, CT: Greenwood Press.

Eagles, Paul F. J., and Susan Muffitt. 1990. "An Analysis of Children's Attitudes toward Animals." *The Journal of Environmental Education* 21 (3): 41–44. doi:10.1080/00958964.1990.10753747.

Freeman, Carrie Packwood. 2010. "Meat's Place on the Campaign Menu: How US Environmental Discourse Negotiates Vegetarianism." *Environmental Communication* 4 (3): 255–76. doi:10.1080/17524032.2010.501998.

Glasman, Laura R., and Dolores Albarracín. 2006. "Forming Attitudes That Predict Future Behavior: A Meta-Analysis of the Attitude–Behavior Relation." *Psychological Bulletin* 132 (5): 778–822. doi:10.1037/0033-2909.132.5.778.

Heberlein, Thomas A. 2012. *Navigating Environmental Attitudes*. New York: Oxford University Press.

Hinton, KaaVonia, and Gail K Dickinson. 2007. *Integrating Multicultural Literature in Libraries and Classrooms in Secondary Schools*. Columbus, OH: Linworth Pub.

Hogan, Patrick Colm. 2009. *Understanding Nationalism: On Narrative, Cognitive Science, and Identity*. Theory and Interpretation of Narrative. Columbus: Ohio State University Press.

Ingram, David. 2000. *Green Screen: Environmentalism and Hollywood Cinema*. Representing American Culture. Exeter: University of Exeter Press.

Jr, George F. McHendry. 2012. "Whale Wars and the Axiomatization of Image Events on the Public Screen." *Environmental Communication* 6 (2): 139–55. doi:10.1080/17524032.2012.662163.

Kalof, Linda, Joe Zammit-Lucia, Jessica Bell, and Gina Granter. 2016. "Fostering Kinship with Animals: Animal Portraiture in Humane Education." *Environmental Education Research* 22 (2): 203–28. doi:10.1080/13504622.2014.999226.

Keen, Suzanne. 2007. *Empathy and the Novel*. Oxford; New York: Oxford University Press.

Kellert, Stephen R., and Miriam O. Westervelt. 1983. *Children's Attitudes, Knowledge and Behaviors toward Animals*. Phase V. Superintendent of Documents, U.S. Government Printing Office, Washington, DC

Kraus, Stephen J. 1995. "Attitudes and the Prediction of Behavior: A Meta-Analysis of the Empirical Literature." *Personality and Social Psychology Bulletin* 21 (1): 58–75. doi:10.1177/0146167295211007.

Louw, Pat. 2006. "Nature Documentaries: Eco-tainment? The Case of MM&M (Mad Mike and Mark)." *Current Writing: Text and Reception in Southern Africa* 18 (1): 146–62. doi:10.1080/1013929X.2006.9678239.

Moore, Ellen Elizabeth. 2016. "Green Screen or Smokescreen? Hollywood's Messages about Nature and the Environment." *Environmental Communication* 10 (5): 539–55. doi:10.1080/17524032.2015.1014391.

Morris, Roy, David B Sachsman, and S. Kittrell Rushing. 2007. *Memory and Myth: The Civil War in Fiction and Film from Uncle Tom's Cabin to Cold Mountain*. West Lafayette, IN: Purdue University Press.

Mossner, Alexa Weik Von, ed. 2014. *Moving Environments: Affect, Emotion, Ecology, and Film*. Ontario: Wilfrid Laurier University Press.

Nussbaum, Martha Craven. 1990. *Love's Knowledge: Essays on Philosophy and Literature*. New York: Oxford University Press.

Oatley, Keith. 2011. *Such Stuff as Dreams: The Psychology of Fiction*. Chichester; Malden, MA: Wiley-Blackwell.

Pinker, Steven. 2011. *The Better Angels of Our Nature: Why Violence Has Declined*. New York: Viking.

Samuels, William Ellery, Lieve Lucia Meers, and Simona Normando. 2016. "Improving Upper Elementary Students' Humane Attitudes and Prosocial Behaviors through an In-Class Humane Education Program." *Anthrozoös* 29 (4): 597–610. doi:10.1080/08927936.2016.1228751.

Schutten, Julie Kalil. 2008. "Chewing on the Grizzly Man: Getting to the Meat of the Matter." *Environmental Communication* 2 (2): 193–211. doi:10.1080/17524030802141752.

Scott, Karen D. 2003. "Popularizing Science and Nature Programming: The Role of 'Spectacle' in Contemporary Wildlife Documentary." *Journal of Popular Film and Television* 31 (1): 29–35. doi:10.1080/01956050309602866.

Smith, Liam David Graham, Sam H. Ham, and Betty Virginia Weiler. 2011. "The Impacts of Profound Wildlife Experiences." *Anthrozoös* 24 (1): 51–64. doi:10.2752/175303711X12923300467366.

Swim, Janet K., and Brittany Bloodhart. 2015. "Portraying the Perils to Polar Bears: The Role of Empathic and Objective Perspective-Taking toward Animals in Climate Change Communication." *Environmental Communication* 9 (4): 446–68. doi:10.1080/17524032.2014.987304.

Waldau, Paul. 2013. *Animal Studies: An Introduction*. New York: Oxford University Press.

Appendices

Appendix 1

Quiz Announcements Used in the Study Described in Chapter 2

Announcement A

An English translation of the original Polish text of the announcement posted on the author's Facebook profile.

"Dear Sirs and Madams, I would like to invite you to an interesting quiz, which will give you the opportunity to read an unpublished fragment of my latest novel *The Lord of the Numbers* and a chance to win a copy of the book. I am currently cooperating with scholars who would like to study the psychological profile of the readers of my novels by using an internet questionnaire. In order to be able to win a copy, you only have to fill out the questionnaire and answer a quiz question. 77 copies of *The Lord of the Numbers* are waiting for the winners! The questionnaire and further information can be found at the website http://badanie-czytelnikow.imas.pl/#welcome [no longer active] All are invited! Marek Krajewski."

Announcement B

The original Polish version of the announcement posted on the author's Facebook profile.

"Szanowni Państwo, chciałbym Państwa zaprosić do ciekawej zabawy, dzięki której będziecie Państwo mogli przeczytać przedpremierowy fragment mojej najnowszej powieści pt. *Władca Liczb*, a także uzyskać szansę wygrania tej książki. Współpracuję obecnie z naukowcami, którzy chcieliby zbadać przy pomocy ankiety internetowej, jaki jest profil psychologiczny czytelników moich powieści. Aby móc wygrać książkę, wystarczy tylko wypełnić tę ankietę i odpowiedzieć na pytanie konkursowe. Na zwycięzców czeka 77 egzemplarzy „Władcy Liczb"! Ankietę oraz dalsze informacje znajdziecie Państwo na stronie http://badanie-czytelnikow.imas.pl/#welcome Zapraszam serdecznie! Marek Krajewski."

Announcement C

An English translation of the original Polish text of the announcement posted on the publisher's Facebook profile.

"Fancy a copy of the latest Book by Marek Krajewski *The Lord of the Numbers*? If so, then now is the only chance to win it – before the official premiere! 77 copies to grab! You just have to take part in an interesting research project – fill out a questionnaire and answer a quiz question. For more information, follow this link -- > http://badanie-czytelnikow. imas.pl/. All are invited!:) ZNAK [the name of the Publisher]"

Announcement D

The original Polish text of the announcement posted on the publisher's Facebook profile

"Macie ochotę na najnowszą książkę Marka Krajewskiego, *Władca liczb*?;) Jeśli tak, tylko teraz jedyna szansa, żeby ją wygrać - jeszcze przed oficjalną premierą! Do zgarnięcia aż 77 egzemplarzy!:) Wystarczy, że weźmiecie udział w ciekawym projekcie badawczym - wypełnicie ankietę internetową i odpowiecie na pytanie konkursowe. Więcej informacji w tym linku -- > http://badanie-czytelnikow.imas.pl/. Serdecznie zapraszamy!:) ZNAK"

Appendix 2

Experimental Narrative Used in the Study Described in Chapter 2

This is an English translation of the original Polish experimental narrative used in the study. The original was later used, and can be found, in Krajewski 2014.

"That I myself was not a follower of Belmispar did not mean that our city was devoid of his acolytes. While I was introducing Leocadie to Zaranek-Plater's mathematical demonology, in one of Wrocław's apartments a likely follower of the Lord of the Numbers was conducting his experiments.

The object of his studies was a monkey named Clotho as a tribute to one of the mythical weavers, who – together with her sisters Lachesis and Atropos – weaved the thread of human existence.

Fifteen years back, that black-and-white capuchin monkey lived in a Venezuelan jungle. The warmth of her mother's belly, to which she clung tightly, filled her with a sense of security when her mother jumped from one woody balsa treetop to another during her air travels. On one of such trips, the monkey felt the muscles of her mother's belly contracting violently. The feeling was unpleasant, and so was the one following it, when the little creature felt the power of gravity for the first time in her life. She turned her head around and did not see the usual sight – tree leaves moving below her. Instead, she noticed brown hairless skin and hard heels stepping firmly on the decomposing undergrowth of the jungle. Intense anxiety made her let go off her mother's belly. She slid down, but did not fall. This was prevented by a net tightly knit from a liana.

On that day, the monkey learned her first lesson about the pain of separation. A native from the tribe of Warao sold her mother to a European trader, and she herself was given to the native's children as a toy. The children would sometimes hug and stroke her, and pricked and pinched her on other times. When, in an act of self-defense, she bit the native's beloved son on his finger, her fate was sealed. First, she was painfully kicked around the house, and then she found herself in the cargo hold of a Dutch freighter going to Amsterdam.

Thus began the true toil of her simian existence. She was sold to an animal wholesaler who then sold her to a travelling circus. There she discovered a new kind of pain – not the slight one which she once felt as a result of venomous ant bites or the pinches, and even kicks, of the native children. No, the new pain was overwhelming and expansive. Its effects were not confined to the particular place on her tiny body touched by the instrument that inflicted suffering: it was spreading constantly and in irregular waves.

The circus trainer was a chain smoker and he liked it that he could do at work what he liked best. First, when the monkey refused to roller skate and walk on stilts, he pulled her by the chain attached to her neck. The animal would fall to the sand of the arena and screamed in fear. Unfortunately, it would then repeat her mistakes and throw away with aversion the many toys and objects it was supposed to use for the amusement of the circus audience. Faced with such failures, it dawned on the circus trainer that he could use a cigarette. He grabbed the monkey by her throat with one hand, and pulled a hood on her head with the other. Convinced that, as a result, the animal would be unable to bite him, he then took a cigarette out of his mouth and pressed it against a tiny heel.

The penetrating shriek of the animal spanned a few registers. The monkey's body, wrapped in an enormous hood, trembled spasmodically, and her nervous system reacted by relaxing sphincters.

The trainer withdrew his hand in disgust, grabbed the creature by its hand, dragged the tiny, still trembling body through the sand of the arena, and then, having waited for an hour, he would scrape off the dust covered muck.

The tiny animal was a fast learner. No more than a week had to pass until the trainer did not have to strain himself anymore and could use cigarettes according to their usual purpose. After a few burns the monkey became obedient. Consistent with the findings of Ivan Pavlov, she associated pain with the darkness of the hood, not with the glow of the cigarette. Whenever darkness would fall upon her, out of fear, she would bite the soft felt of the hood. Eventually, its mere sight was enough to make her do what was wanted of her. At night, when it would become pitch black, the monkey would go insane and bite everything around her.

Soon she became an attraction of the travelling circus. She let herself be stroked and fed. She even learned how to shake hands with the audience. One day the circus found itself in the distant Poland, where it amused German soldiers. In Cracow, she became the darling of the daughter of a certain general. During one of the performances, the child was encouraged by a clown to shake hands with the animal. She did so and precisely at this moment all lights went out. It was obvious for the monkey that the hood had once again separated her from light. She reacted the way she usually did when it would become dark. She bared her teeth and made use of them.

On the very same night, the monkey was sold for pennies to an organ grinder, who did not want anything from her apart from sitting on his organ. He looked after her with such care that he did not even economize on lamp oil and put a lit lamp next to her cage at night. Unfortunately, that *dolce vita* did not last long. The organ grinder passed away, and the monkey, christened Clotho by her new owner, only then encountered the true reality of pain.

This one was the most terrifying the animal ever felt. It twisted and paralyzed her body for a few seconds. Its source was neither a particular point or area on her body – the pain now resided at the very center of the capuchin, it wrenched her with spasms and threw her against the walls of the cage.

The worst thing was that she could do nothing about it. The man would put inside the cage an iron stand with two ladders leading to a small platform. One of them was black, the other white. Lying on the platform, there was a walnut. The creature would happily climb for the walnut – using either the white or the black ladder. Then the man would draw out two protruding wires in her direction. Electricity would twist her body and force a high-pitched shriek out of the tiny throat. The man would smile friendly, say something in a silent voice and touch one or the other ladder with a pointer – the white and the black one, in turns. Clotho did not know what was on her tormentor's mind. Afraid of the wires, she jumped from one ladder to the other like crazy, blindly. Then the man would apply electric shocks again. Apparently, he demanded something else. She did not understand that he wanted to make her disorderly jumps less chaotic – that all that he wanted was that she first climbed the black ladder, and then immediately the white one.

Clotho failed to grasp the man's intentions. She was helpless. All she could do was to look into the eyes of the tormentor approaching her. And then to suffer."

Appendix 3
Control Narrative Used in the Study Described in Chapter 2

This is an English translation of the original Polish control narrative used in the study. The original was later used, and can be found, in Krajewski 2014.

"The man standing just outside my garden kept his eye on me constantly. When I took a look at him, he raised his bowler hat, revealing a few streaks of hair stuck firmly to his bald skull. I responded with a similar gesture, the only difference being that my head was entirely bald as I had nothing but utter scorn for all kinds of comb-overs. The heat of July morning obviously made no impression on the corpulent gentleman, since he was dressed in an unusually hermetic way, so to speak. He wore a brown three-piece suit made of wool and a bowler hat of the same color. A bow tie of a slightly lighter shade of brown sealed his outfit under his chin, preventing any influx of fresh air from above. In his browns, he looked like an English lord who by some strange twist of fate had been transported to the white-hot Wrocław from his moorlands in Yorkshire.

'Do I have the honor of speaking with Mr. Edward Popielski?' he asked.

'Yes indeed, it's me.' I replied.

I scanned the man from head to toes, and could not resist bursting with laughter upon seeing his Georgian elegance, which was complemented by knickerbockers, the bottoms of which were tucked in long checkered gaiters.

'I have no idea what exactly you find so funny about me, but I wanted to assure you I'm not a figure from a cabaret, Mr. Popielski.' The man, who on first look seemed to be a sexagenarian, raised his voice. 'I am Count Władysław Zaranek-Plater.'

He raised his eyes to the sky. Apparently, he was expecting an apology and an expression of reverence, but what he got was something else entirely.

'I'm about to start cheering', I smiled broadly. 'And then I'll ask you to sign my friendship book...'

That joke apparently made a bigger impression on him than the heat, because he took off the bowler-hat and wiped sweat of his forehead with a large checkered handkerchief, going through amazing contortions so as not to ruin his elaborate hairdo.

'I am a friend of your boss, Aleksander Beck.' He handed me his name card and wheezed scornfully, putting his bowler-hat back on his head. 'I know what you really do for a living and I wanted to hire you to do a certain important and exceptionally well-paid job.'

The adverb "really" put me out of the mood for making jokes. My stint as a "legal representative" at Beck's law firm, which had lasted four years already, was no mystery and my daily job no cause for sensation. If anyone had followed me, they would have seen that every day at eight in the morning I would come to the office at Świerczewskiego Street, browse the files of current cases and make some notes. If that hypothetical investigator had felt like observing me a bit longer, then he would have learned that I met with various people and that I just talked to them.

What I "really" did, however, was not that far from my previous job as a detective. For the so-called gathering of evidence consisted in looking for vices and weaknesses of the prosecution witnesses and then skillfully using those flaws. Since this activity often involved a bit of delicate blackmail – if not intimidation – I preferred that third parties, such as this count straight out of an operetta, did not know what my job "really" consists in.

'Please do not bother people in the street', I said in a calm voice. 'If you are a salesman, sell your whetstones to somebody else.'

With a volume of Cicero tucked under my arm, I walked off toward the ruins at the corner of Górnickiego and Benedektyńska street. The cathedral towers were gleaming in the distance.

Count Zaranek-Plater was running after me.

'Please wait for me, Mr. Popielski.' He breathed heavily, and the tone of his voice seemed pleading to me. 'There's a cab waiting for us just around the corner. We can go to Mr. Beck, and he will confirm everything: that I talked to him about you, and that he recommended you to me... At this very moment he is on the beach by Odra with his family, but he will not be upset if we disturb him ... He is an old friend, a compatriot, we come from the same place.'

I stopped. I knew that on that Sunday Beck was in fact going with his wife and two children to the beach by Odra. Just yesterday, his beautiful wife visited our office and – disturbing with her graceful moves the peace of my young assistant – she interrogated me thoroughly (as if I had been some kind of expert on water engineering!) about the dangers lurking for those bathing in the vicinity of the Opatowicka Island.

'All right.' I watched with amusement as Zaranek-Plater tried to evenly spread the streaks of hair on his skull. 'I believe your Lordship... Well ... I will not invite you into my house as my cousin is giving lessons at the moment. I suggest that we take a walk in the Botanical Garden. This is where I have just been heading and this is also where your Lordship will present his job order to me. So? Shall we take a walk?'

'We will take a ride! This is a hired driver.' The count replied with confidence and pointed toward a cab parked on the corner of Ładna and Miła, with the driver standing next to it, sporting a wife-beater and a beret with a stem on top. And we will not go to the Botanical Garden, but to the Main Train Station.

'What for? Do you want to take me out on a suburban trip?'

'I lead a very regular life! Eleven o'clock is fast approaching and at this time of the day I always eat a cream-filled biscuit and drink coffee at the Main Train Station café... I invite you to a biscuit!'

I felt that I, too, was beginning to be affected by the July heat. I took off my hat, unbuttoned another button of my shirt, whose collar I had previously laid out on the shoulders of my jacket.

'Your Lordship' I waved myself with the hat. 'I suffer greatly when I am forced to waste my time. And I do not like to suffer. How can I know that I will like your biscuit? I am not looking for an extra pay.'

'Mr. Popielski', Zaranek-Plater took me by the elbow and started to lead gently toward the cab. 'If you reject my offer, I will give you five hundred for the trouble and the cab will drive you back home.'

'And if I take it?'

'Then my friend Olek Beck will curse me because he will lose a reliable employee!'

'Could your Lordship be a bit clearer?' Now it was my turn to wipe sweat off my bald scalp. 'It is too hot for conditionals without consequents. What will happen if I take your job offer? What will happen, my Lord? Finish that sentence!'

'You will not have to work until the end of your life', said the count lazily and whispered a very large and tempting sum right into my ear. 'So what do you think? Will you take the risk and spend a moment in my company? The worst-case scenario is that you will then show the five hundred to your cousin and say: "I'm sorry I'm late for lunch. I wasted some time with a certain eccentric!"' I nodded my head, and after a short while, I was rocking on the back seat of a Warszawa and looking with disgust at the gray-bristle covered neck of the cab driver."

Appendix 4

Questionnaire Used in the Study Described in Chapter 2

Questionnaire A

An English translation of the original Polish questionnaire used in the study

Introduction

Study of the psychological profile of Marek Krajewski's readers

Thank you for your interest in taking part in our study. The Questionnaire consists of several parts. First, you will read a fragment of the latest book by the author, and then we will ask you a series of questions. We hope that filling out the questionnaire will be an interesting experience for you.

The questionnaire is anonymous. All answers and data gathered are confidential. They will be used collectively and only for research purposes.

The study should take ca. 20–25 minutes.

Read closely the following fragment of the latest book by Marek Krajewski. Then turn to the questionnaire. Some of the questions will concern your general views, others your impressions about the text.

[Participants were then randomly assigned either to Experimental Narrative or Control Narrative]

Main Part

The purpose of this study is to study the views and attitudes of Marek Krajewski's readers. You will see a questionnaire consisting of 40 questions. Read closely each of them and mark to what degree you agree with each of them.

There are no good and bad answers here. We will appreciate honest answers.

PS. At one spot we placed a control question, where you will be asked to mark one particular answer indicated by us. Don't miss it, good luck!

Scale of answer choices:

Completely disagree 1
Disagree 2
Somewhat disagree 3
Neither agree nor disagree 4

Somewhat agree 5
Agree 6
Completely agree 7

Page 1

1 I see myself as extroverted, enthusiastic.
2 Education in the humanities gives one as good prospects as education in STEM.
3 In vitro fertilization is immoral.
4 I see myself as critical, quarrelsome.
5 Our country needs a powerful leader, in order to destroy the radical and immoral currents prevailing in society today.
6 Genetically modified food should be freely sold in stores.
7 The slaughter of whales and dolphins should be immediately stopped, even if it means that some people will be put out of work
8 Our country needs free thinkers who will have the courage to stand up against traditional ways, even if this upsets many people.
9 I see myself as dependable, self-disciplined.
10 The "old-fashioned ways" and "old-fashioned values" still show the best way to live.

Page 2

11 Our society would be better off if we showed tolerance and understanding for untraditional values and opinions.
12 Polish citizens should have more access to guns than they have today.
13 The suffering of animals is an acceptable price for inventing drugs for humans.
14 I see myself as anxious, easily upset.
15 God's laws about abortion, pornography and marriage must be strictly followed before it is too late, violations must be punished.
16 Cultural minorities should be supported and protected.
17 I support the legalization of marijuana.
18 Human needs should always come before the needs of animals.
19 I see myself as open to new experiences, complex.
20 The society needs to show openness towards people thinking differently, rather than a strong leader, the world is not particularly evil or dangerous.

Page 3

21 It would be best if newspapers were censored so that people would not be able to get hold of destructive and disgusting material.
22 Many good people challenge the state, criticize the church and ignore "the normal way of living."

23 I see myself as reserved, quiet.

24 Our forefathers ought to be honored more for the way they have built our society, at the same time we ought to put an end to those forces destroying it.

25 The health care system should be privatized.

26 I feel personally responsible for helping animals in need.

27 People ought to put less attention to the Bible and religion; instead they ought to develop their own moral standards.
   • Control question. Tick „somewhat disagree"

28 I see myself as sympathetic, warm.

29 There are many radical, immoral people trying to ruin things; the society ought to stop them.

30 I would like EURO currency to be introduced in Poland instead of the Polish Złoty.

Page 4

31 The low costs of food production do not justify maintaining animals under poor conditions.

32 I see myself as disorganized, careless.

33 It is better to accept bad literature than to censor it.

34 Facts show that we have to be harder against crime and sexual immorality, in order to uphold law and order.

35 Apes should be granted rights similar to human rights.

36 I see myself as calm, emotionally stable.

37 The situation in the society of today would be improved if troublemakers were treated with reason and humanity.

38 I see myself as conventional, uncreative.

39 Basically, humans have the right to use animals as we see fit.

40 If the society so wants, it is the duty of every true citizen to help eliminate the evil that poisons our country from within.

Questions Concerning the Text

Now we would like to ask you to answer questions concerning the fragment of Marek Krajewski's novel you read in the beginning. Please mark the number you think best fits your opinion about the text.
   Scale of answer choices:

Completely disagree 1
Disagree 2
Somewhat disagree 3
Neither agree nor disagree 4
Somewhat agree 5
Agree 6
Completely agree 7

[Version for the experimental group:]

 1 While I was reading the narrative, I could easily picture the events in it taking place.
 2 While I was reading the narrative, activity going on in the room around me was on my mind.
 3 I could picture myself in the scene of the events described in the narrative.
 4 I was mentally involved in the narrative while reading it.
 5 After the narrative ended, I found it easy to put it out of my mind.
 6 I wanted to learn how the narrative ended.
 7 The narrative affected me emotionally.
 8 I found myself thinking of ways the narrative could have turned out differently.
 9 I found my mind wandering while reading the narrative.
10 The events in the narrative are relevant to my everyday life.
11 The events in the narrative have changed my life.
12 I had a vivid mental image of the monkey.
13 I had a vivid mental image of the circus.

[Version for the control group:]

 1 While I was reading the narrative, I could easily picture the events in it taking place.
 2 While I was reading the narrative, activity going on in the room around me was on my mind.
 3 I could picture myself in the scene of the events described in the narrative.
 4 I was mentally involved in the narrative while reading it.
 5 After the narrative ended, I found it easy to put it out of my mind.
 6 I wanted to learn how the narrative ended.
 7 The narrative affected me emotionally.
 8 I found myself thinking of ways the narrative could have turned out differently.
 9 I found my mind wandering while reading the narrative.
10 The events in the narrative are relevant to my everyday life.
11 The events in the narrative have changed my life.
12 I had a vivid mental image of the restaurant.
13 I had a vivid mental image of the taxi driver.

Demographic Data Questions

Thank you for all your answers. Now there are only a few demographic questions left.

Mark the size of the town you currently live in:

1 Village
2 Town 25,000 and fewer inhabitants
3 Town 25–50 tys. inhabitants
4 Town 51–100 tys. inhabitants
5 Town 101–200 tys. inhabitants
6 City 201–500 tys. inhabitants
7 City above 500,000 inhabitants

What is your educational background:

1 Primary/ junior high/vocational
2 High school/ post-high school
3 Ba./Ma./Ph.D

Answer yes or no:

Do you have siblings?
Do you have a car?
Do you keep pets?
Do you exercise regularly?
Do you abstain from drinking alcohol?
Are you a vegetarian?
Are you a vegan?
Is crime fiction your FAVORITE literary genre?
Are you generally in good health?
Have you ever suffered any injuries?

If you have siblings:
    State the age and gender of your siblings
    ....
    If you have a car:
    What brand of car do you have and how long you have had it for?
    ....
    If you have pets:
    Which species and for how long you have had them for?
    ....
    If you exercise regularly:
    What kind of exercises and for how long have been exercising?
    ....
    If you have been injured:
    What was the cause?

1 Transport accident
2 Chemical substances

3 Mechanical injury
4 Animal attack
5 Other, which? ....

Quiz

That is all, thank you for your efforts! Now let us have some fun :)

To take part in the quiz, please fill out the form below and answer a quiz question

Your e-mail address: ....

Name and address: ....

[] I hereby confirm that I have read and accept the terms of the quiz „The study of the psychological profile of Marek Krajewski's readers " and that I agree to the processing of my personal data contained in the form, by IMAS International Sp. z o.o. Wrocław, ul. Braci Gierymskich 156, only for the purpose and to extend necessary for running the quiz – not longer than for the duration of the quiz and until any potential demands are stale.

The quiz question is:

How many have fell victim to the Lord of Numbers in the book? 5, 6, 8, 9, 16, 17?

Thank you for participating in the study!

Questionnaire B

The original Polish version of the questionnaire used in the study

WSTĘP

Badanie profilu czytelników Marka Krajewskiego!

Dziękujemy Ci za chęć uczestnictwa w naszym badaniu. Ankieta składa się z kilku części, w pierwszej z nich przeczytasz fragment najnowszej książki Autora, a następnie zadamy Ci szereg pytań. Mamy nadzieję, że wypełnianie ankiety będzie dla Ciebie ciekawym przeżyciem.

Ankieta jest anonimowa. Wszystkie odpowiedzi oraz dane są poufne i będą wykorzystywane jedynie zbiorczo, w celach badawczych.

Ankieta potrwa ok. 20–25 minut.

Fragment Książki

Przeczytaj uważnie fragment najnowszej książki Marka Krajewskiego. Następnie przejdź do kwestionariusza. Część pytań dotyczyć będzie Twoich ogólnych poglądów, część zaś Twoich wrażeń z przeczytanego tekstu.

Część Zasadnicza

Celem tego badania jest określenie, jakie poglądy i postawy mają czytelnicy Marka Krajewskiego. Zobaczysz kwestionariusz, który składa się z 40 stwierdzeń. Przeczytaj każde z nich uważnie i zaznacz przy poszczególnych stwierdzeniach, do jakiego stopnia zgadzasz się lub nie zgadzasz z każdym z nich.

Nie ma tu dobrych ani złych odpowiedzi, prosimy o szczere odpowiedzi.

PS.W pewnym miejscu umieściliśmy stwierdzenie kontrolne, w którym prosimy o zaznaczenie jednej, wskazanej przez nas odpowiedzi. Nie przeocz go, powodzenia!

Skala odpowiedzi dla wszystkich stwierdzeń na stronach 1–4:

Zdecydowanie się nie zgadzam |1
Nie zgadzam się |2
Raczej się nie zgadzam |3
Ani się zgadzam, ani nie zgadzam |4
Raczej się zgadzam |5
Zgadzam się |6
Zdecydowanie się zgadzam |7

Strona 1

1 Postrzegam siebie jako osobę lubiącą towarzystwo innych, aktywną i optymistyczną.
2 Kierunki humanistyczne dają równie dobre perspektywy, jak kierunki techniczne.
3 Zapłodnienie metodą in vitro jest niemoralne.
4 Postrzegam siebie jako osobę krytyczną względem innych, konfliktową.
5 Nasz kraj potrzebuje silnego przywódcy, po to by położyć kres radykalnym i niemoralnym prądom, które przeważają dziś w społeczeństwie.
6 Żywność genetycznie modyfikowana powinna być dopuszczona do obrotu handlowego.
7 Ubój delfinów i wielorybów powinien być natychmiast wstrzymany, nawet jeśli oznaczałoby to utratę pracy przez niektórych ludzi.
8 Nasz kraj potrzebuje wolnomyślicieli, którzy będą mieli odwagę przeciwstawić się tradycyjnym obyczajom, nawet jeśli nie spodoba się to wielu ludziom.
9 Postrzegam siebie jako osobę sumienną, zdyscyplinowaną.
10 „Staroświeckie obyczaje" i „staroświeckie wartości" wciąż pokazują najlepszy sposób życia.

Strona 2

11 Nasze społeczeństwo miałoby się lepiej, gdybyśmy okazywali tolerancję i zrozumienie dla nietradycyjnych wartości i opinii.
12 Polscy obywatele powinni mieć większy dostęp do broni niż obecnie.
13 Cierpienie zwierząt jest dopuszczalną ceną za wynajdywanie leków dla ludzi.
14 Postrzegam siebie jako osobę pełną niepokoju, łatwo wpadającą w przygnębienie.
15 Boskie prawa dotyczące aborcji, pornografii i małżeństwa muszą być ściśle przestrzegane, zanim będzie za późno, a ich naruszenia powinny być karane.
16 Powinno się wspierać i chronić mniejszości kulturowe.
17 Jestem za legalizacją miękkich narkotyków.
18 Potrzeby ludzkie zawsze powinny być ważniejsze od potrzeb zwierząt.
19 Postrzegam siebie jako osobę otwartą na nowe doznania, w złożony sposób postrzegającą świat.
20 Bardziej niż silnego przywódcy społeczeństwo potrzebuje otwartości wobec ludzi myślących inaczej – świat nie jest szczególnie zły lub niebezpieczny.

Strona 3

21 Byłoby najlepiej, gdyby gazety były cenzurowane w celu uniemoż-
liwienia ludziom kontaktu ze szkodliwymi czy też odpychającymi
treściami.

22 Wielu dobrych ludzi przeciwstawia się państwu, krytykuje Kościół i
ignoruje „normalny sposób życia".

23 Postrzegam siebie jako osobę zamkniętą w sobie, wycofaną i cichą.

24 Naszych przodków powinno się bardziej doceniać za to, jak zbudow-
ali nasze społeczeństwo, z drugiej strony zaś powinniśmy położyć
kres siłom, które to społeczeństwo niszczą.

25 Służba zdrowia powinna być sprywatyzowana.

26 Czuję się osobiście odpowiedzialny(a) za pomoc potrzebującym
zwierzętom

27 Ludzie powinni przywiązywać mniej wagi do Biblii czy religii, a
zamiast tego powinni wynajdywać swoje własne standardy moralne.
• Pozycja kontrolna. Zaznacz odpowiedź „Raczej się zgadzam"

28 Postrzegam siebie jako osobę zgodną, życzliwą.

29 Jest wielu radykalnych, niemoralnych ludzi próbujących wszystko
zrujnować – społeczeństwo powinno ich powstrzymać.

30 Chciałbym aby w Polsce wprowadzono EURO zamiast złotówki.

Strona 4

31 Niskie koszty produkcji pożywienia nie uzasadniają hodowania
zwierząt w złych warunkach.

32 Postrzegam siebie jako osobę źle zorganizowaną, niedbałą.

33 Sądzę, że lepiej jest zaakceptować niemoralną literaturę, niż ją
cenzurować.

34 Fakty pokazują, że po to by utrzymać prawo i porządek, powinniśmy być ostrzejsi wobec przestępczości oraz wobec niemoralności
seksualnej.

35 Uważam, że małpom człekokształtnym powinno się przyznać prawa
podobne do praw człowieka.

36 Postrzegam siebie jako osobę niemartwiącą się, stabilną
emocjonalnie.

37 Sytuacja dzisiejszego społeczeństwa poprawiłaby się, gdyby ludzie
sprawiający kłopoty byli traktowani z rozsądkiem i po ludzku.

38 Postrzegam siebie jako osobę trzymającą się schematów, biorącą
rzeczy wprost.

39 Ludzie mają prawo posługiwać się zwierzętami wedle swego
uznania.

40 Jeśli społeczeństwo tak chce, to obowiązkiem każdego prawdzi-
wego obywatela jest pomóc wyplenić zło, które zatruwa nasz kraj
od środka.

Pytania Do Tekstu

Teraz prosimy o odpowiedź na pytania dotyczące przeczytanego przez Ciebie na początku fragmentu powieści Marka Krajewskiego. Proszę zaznacz liczbę, która najlepiej odpowiada Twojej opinii na temat tego tekstu. Skala odpowiedzi dla wszystkich stwierdzeń:

Zdecydowanie się nie zgadzam |1
Nie zgadzam się |2
Raczej się nie zgadzam |3
Ani się zgadzam, ani nie zgadzam |4
Raczej się zgadzam |5
Zgadzam się |6
Zdecydowanie się zgadzam |7

[Wersja dla grupy eksperymentalnej:]

1 Kiedy czytałem tekst, łatwo mi było wyobrazić sobie wydarzenia, które były w nim opisywane.
2 Kiedy czytałem tekst, zwracałem uwagę na to, co działo się w pomieszczeniu, w którym akurat się znajdowałem.
3 Mogłem wyobrazić sobie samego siebie w miejscu wydarzeń opisanych w tekście.
4 Byłem zaangażowany myślami w tekst, kiedy go czytałem.
5 Kiedy skończyłem czytać tekst, łatwo mi było przestać o nim myśleć.
6 Chciałem się dowiedzieć, jak tekst się skończy.
7 Tekst poruszył moje emocje.
8 Zastanawiałem się nad tym, jak inaczej mógłby się skończyć ten tekst.
9 Kiedy czytałem tekst, myślami byłem gdzie indziej.
10 Wydarzenia przedstawione w tekście są istotne z punktu widzenia mojego codziennego życia.
11 Wydarzenia przedstawione w tekście zmieniły moje życie.
12 Czytając tekst miałem przed oczyma wyraźny obraz małpki.
13 Czytając tekst miałem przed oczyma wyraźny obraz cyrku.

Metryczka

Dziękujemy za wszystkie odpowiedzi. Pozostało jeszcze kilka pytań metryczkowych.

Zaznacz wielkość miejscowości w której aktualnie (głównie) mieszkasz.

1 Wieś
2 Miasto do 25 tys. mieszkańców
3 Miasto 25–50 tys. mieszkańców

4 Miasto 51–100 tys. mieszkańców
5 Miasto 101–200 tys. mieszkańców
6 Miasto 201–500 tys. mieszkańców
7 Miasto powyżej 500 tys. mieszkańców

Jakie jest Twoje wykształcenie?

1 Podstawowe/gimnazjum/ zawodowe
2 Średnie/ policealne
3 Wyższe/Licencjat

Zaznacz TAK lub NIE na poniższe pytania. Czy...?

Posiadasz rodzeństwo
Posiadasz samochód
Posiadasz zwierzęta domowe
Uprawiasz regularnie sport
Jesteś abstynentem
Jesteś wegetarianinem
Jesteś weganinem
Kryminały to mój NAJBARDZIEJ ulubiony gatunek literacki
Ogólnie rzecz biorąc, cieszę się dobrym zdrowiem
Czy kiedykolwiek doznałeś jakichś dotkliwych urazów fizycznych?

Jeśli masz rodzeństwo:
   Podaj proszę płeć i wiek rodzeństwa (jeśli masz kilkoro rodzeństwa, wymień po przecinku)
   ....
   Jeśli masz samochód:
   Podaj proszę markę samochodu i od jak długo go posiadasz
   ....
   Jeśli masz zwierzęta:
   Podaj proszę jakie zwierzęta posiadasz i od jakiego czasu
   ....
   Jeśli uprawiasz sport:
   Jaki sport uprawiasz i od jakiego czasu?
   ....
   Jeśli miałeś/aś wypadek:
   W wyniku czego doznałeś(aś) dotkliwych obrażeń fizycznych?

1 wypadek lokomocyjny
2 działanie substancji chemicznych
3 działnie maszyn
4 atak zwierząt
5 inne, jakie? ....

Konkurs

To już wszystkie pytania, dziękujemy za Twój wysiłek! Teraz przejdźmy do przyjemności:)

Aby wziąć udział w konkursie, prosimy Cię o wypełnienie poniższego formularza i odpowiedź na pytanie konkursowe.

Twój adres e-mail: ....

Imię i nazwisko: ....

[] Oświadczam, że zapoznałem/am się i akceptuję Regulamin konkursu „Badanie profilu psychologicznego czytelników Marka Krajewskiego " oraz wyrażam zgodę na przetwarzanie danych osobowych, zawartych w formularzu zgłoszeniowym, przez IMAS International Sp. z o.o. z siedzibą we Wrocławiu, ul. Braci Gierymskich 156, wyłącznie w celu i zakresie niezbędnym dla przeprowadzenia konkursu - nie dłużej niż przez okres przeprowadzania konkursu i do przedawnienia ewentualnych roszczeń.

Pytanie konkursowe brzmi:

Ile ofiar pochłonął tytułowy Władca Liczb? 5, 6, 8, 9, 16, 17?
Dziękujemy za udział w ankiecie!

Index

Note: Italic page numbers refer to figures & page numbers followed by "n" denote endnotes.

Aaltola, Elisa 2, 137
AAS *see* Animal Attitudes Scale (AAS)
Abate, Michelle Ann 6
Abrams, M. H. 105n2
abstractions 119–24;
Actman, Jani 3
affirmative action 93
Ajzen, Icek 158n3
Albarracín, Dolores 158n3
Alsbrook, E. Y. 151n2
American Humane Association 10
Ames, Daniel L. 102–3
"Am I Blue?" (Walker) 15, 43, 96, 121, 123, 133; arguments 88–93; experiment and results 93–5
Ammons, Elizabeth 8, 9, 156
anecdotal evidence 3, 66n2
animal abuse 2, 3, 70
animal advocates 1–4, 10, 25, 88
Animal Attitudes Scale (AAS) 27–9, 31, 47n2, 47n3
animal cruelty, in childhood 10–11
animal ethics 4, 5
animal liberation movement 91
animal oppression 38
animal studies 1, 6; literary 5, 43
animal welfare 34, 123; attitude toward *see* attitude toward animal welfare; campaigns 39; farm 74; legislation 4; policies, practical implications for 129; pro-animal welfare 27, 63, 64, 78–9, 94; public's concern about 82; reforms 37
anomalous replotting 131
Another Part of the City (McBain) 78
ANOVA 48n13
anthrozoology 1

Appel, Markus 147, 148
Arendt, Florian 158n2
Argentino, Concetta M. 71
argumentation 2, 88–90
Aristotle 7
Arluke, Arnold: *Considering Animals* 4–5
Ars Poetica (Horace) 6
Ascione, Frank R. 157n1
ATAW Scale *see* Attitudes Toward Animal Welfare (ATAW) Scale
ATDS *see* Attitudes Toward Dogs Scale (ATDS)
ATHS *see* Attitudes Toward Horses Scale (ATHS)
attitude 26–9, 30, 35, 53; definition of 26; toward animal welfare *see* attitude toward animal welfare
Attitudes Toward Animal Welfare (ATAW) Scale 31–2, 47n4, 47n5, 62–3, 71, 121, 141, 150
Attitudes Toward Dogs Scale (ATDS) 121–3
Attitudes Toward Horses Scale (ATHS) 121, 122
attitude toward animal welfare 27, 28, 31–2, 39, 45, 70, 79, 94, 101, 114, 116–18, 138, 142, 145, 154; animal stories' influence on 40–1, 63–4, *64*, 150; fictional narratives, impact of 147; perceived fictionality of the text, impact of 77; pro-animal welfare attitude 27, 63, 64, 78–9, 94, 105, 116, 147; *see also* animal welfare
attitudinal impact over time 145–51
Attridge, Derek 35
Austen, Jane 87
Ayala, Francisco José 59, 116

Babe 3–4, 34, 59, 113, 120, 157
Baeten, Jan 153
Bagust, Phil 158n2
Bakker, Marjan 46
Bal, P. Matthijs 9, 39, 42, 133, 148
Bambi 82
Barcz, Anna 5, 6
Barthes, Roland 41, 61
Batson, C. Daniel 64, 65, 70, 79, 114, 121–2, 158n3
Beastly Things (Leon) 38–9
Beautiful Joe (Saunders) 15, 42, 88, 101–5, 122, 123, 155; experiment and results 105; first-person narration 101–4
Becker, Kate 25
Bekoff, Marc 17n7
Beloved (Morrison) 7, 156
Berger, Ronald J. 17n5
Berns, Gregory S. 151
bestselling authors 54, 57
Betchart, Nancy S. 27, 59
Between the World and Me (Coates) 89, 90
Birke, Dorothee 101
Black Beauty (Sewell) 1, 9, 34, 37, 42, 45, 54, 59, 70, 88, 101, 102, 104–5, 113, 120, 146–7, 154, 155, 157
Bloodhart, Brittany 158n2
Bloom, Harold 74, 75
Boehrer, Bruce Thomas 117
Bogart, Karen 71.
Boggs, Colleen Glenney 3, 5, 58, 88
Bona, Mary Jo 6
Booth, Wayne C. 6–7, 101, 156
Bough, Jill 113
Boxenbaum, Harold 119
Boyd, Brian 81
Bożek, Jakub 82n2
Braunsberger, Karin 158n4
Breaking Bad (Gilligan) 16, 85–7, 128
Breier, Davida Gypsy 43, 73
Brock, T. C. 64, 70, 80, 129–34, 138; on thematic relation between texts and beliefs 134; "The Role of Transportation in the Persuasiveness of Public Narratives," 130–2; Transportation Scale 32, 133
Brockway, Laurie Sue 110–11
Brownell, Ginanne 56
Brunetti, Commissario Guido 38
Brunker, Mike 95
Buell, Lawrence 9
Buettner, Angi 38

Burda, Katarzyna 15
Burke, Michael 40
Byrd, Robert 74

Caesar, Terry 66n1
Cameron, W. Bruce 66n1; *A Dog's Purpose* 54
Camfield, Gregg 3
Carey, John 134
Castricano, Jodey 30
cause-and-effect relation 11
Cela-Conde, Camilo J. 59, 116
Chaiken, Shelly 26
Chao, Melody Manchi 106n5
Chevalier, Judith A. 54
Chez, Keridiana 104
Chrulew, Matthew 6
Clarissa (Richardson) 8, 9
Clark, Anna 7
Coates, Ta-Nehisi 89; *Between the World and Me* 89, 90
Cohen, Barry H. 12
Cohen, Jacob 45
Coleman, Zendaya: *Spiderman: Homecoming* 112
color-blind casting 112
The Color Purple (Walker) 7, 90–3, 156
Considering Animals (Arluke and Sanders) 4–5
Considering the Horse (Rashid) 138–9
controlled experiment 11–13, 42
Cook, Michael 153
Corbey, Raymond 2, 10
Corman, Lauren 30
Cornelisse, Tara M. 158n4
correlational studies 11
Cosslett, Tess 3, 88, 104
Couch, Aaron 86, 87
Cowen, Richard 118
Crano, William D. 26, 95
Crime and Punishment (Dostoyevsky) 16, 43, 95–101; emotional fragility 97–9; experiment and results 99–101; moral judgments, conveying 95–7
criminology 129
Cronon, William 17n5
cruel behavior 10–11

Danna, C. L. 34
Daudet, Lucien 36
da Vinci, Leonardo: "Last Supper," 48n8
Davis, Janet M. 3, 17n1, 43, 104, 155

Dawkins, Marian Stamp 2
"The Dead Body and the Living
    Brain" (Fallaci) 43; biases 114–15;
    experiment and results 115–18;
Death in Breslau (Krajewski) 58
The Death to Art (Gnedov) 35
De Botton, Alain 36
debriefing 33
Deckha, Maneesha 38
Delamater, Jerome 153
DeMello, Margo 5, 17n2, 124n1
Demir, Nilüfer 136
Demme, Jonathan: *Silence of the
    Lambs, The* 66n3
Derrida, Jacques 17n3, 25, 64
detective fiction 24, 58, 61, 112
detectives 58–62
detective stories 128, 153
Dewey, John 35
Di Bella, Maria Pia 129
Dickens, Charles 7
Dickinson, Gail K. 7, 156
Dickstein, Morris 88
Disney: *Bambi* 82
A Dog's Purpose (Cameron) 54
Dorré, Gina M. 17n2
Dostoyevsky, Fyodor 120, 123, 134;
    Crime and Punishment 16, 43,
    95–101
double-blind studies 12
Doyle, Sir Arthur Conan 112; *The
    Hound of the Baskervilles* 58
dreaded comparison 91
D'Souza, Richard W. 119
Duchamp, Marcel 48n8
Dumezweni, Noma 112
Dunning, Thad 43, 55–6
Dupin, C. Auguste 24, 58

Eagles, Paul F. J. 157n1
Eagly, Alice Hendrickson 26
Earthsea (Le Guin) 111–12, 113
ecocriticism 1, 4–6
ecological validity 13, 43, 55, 149
Eisenman, Stephen 39
Eisnitz, Gail A. 3, 120, 141, 148;
    Slaughterhouse 43, 73–6, 79, 99,
    113, 138–40
Eitler, Pascal 3
Elementary (television series) 112–13
Elick, Catherine L. 88
Elist, Jasmine 38
Elkins, James 129
Ellithorpe, Morgan E. 151n2

Elmer, G. I. 34
Elżanowski, Andrzej 2
Empathy and the Novel (Keen) 17n9
empirical evidence, importance
    of 10–11
environmental humanities 14
environmental imagination 9
equine welfare 1
The Ethical Journalism Network 136
Ewoldsen, David R. 151n2
experimental social science 14
experimental story 32, 80, 132, 139
experimenter effects 71
extinction studies 4–6
Eysenck, Michael W. 55

Faktorovich, Anna 57
Fallaci, Oriana 43, 115, 119–20,
    123, 133; "The Dead Body and the
    Living Brain," 43, 114–19
Farm Animal Rights Movement 38
Farner, Geir 106n3
Farrimond, Hannah 16
Faye, Lyndsay 113
fear motivation theory 139, 141
Felski, Rita 15, 25
Fiamengo, Janice Anne 43
fiction 7; animal narratives as 70;
    crime 53; detective 24, 58, 61,
    112; persuasive influence of 148;
    recognition of 72–3; science 111;
    vs. truth 78–82
Fifty Shades of Grey (James) 16,
    110–13
first-person narration 101–4, 113
    see also third-person narration
Fiske, Susan 114
Flood, Alison 6
Flynn, Clifton P. 29
Foakes, R. A. 134
Foer, Jonathan Safran 2
Foley, Robert 142n5
Foster, Jodie 58
Foulkes, A. Peter 61, 88
Francione, Gary L. 4, 10, 17n7, 30
François, Louis 8
Frank, David A. 137
Freeman, Carrie Packwood 158n4
frog-human scale 116, 117, 124n4
Fugitive Slave Act 8

Garrard, Greg 17n2
Garrett, Brandon L. 137
Genette, Gérard 76

Genosko, Gary 119
Gerson, Carole 104
Gilligan, Vince 86; *Breaking Bad* 16, 85–7, 128
Gilson, David 5
Giovannini, Dino 149
Glasman, Laura R. 158n3
Gnedov, Vasilisk: *The Death to Art* 35
Goldman, Alan H. 35
Gone with the Wind 110
Gottschall, Jonathan 135–6; *The Storytelling Animal: How Stories Make Us Human* 135–6
Grayson, Stephanie 47n2, 47n3
Greenfield, Nicole 3
Green, M. C. 64, 70, 80, 129–34, 138; on relation between texts and beliefs 134; "The Role of Transportation in the Persuasiveness of Public Narratives," 130–2; Transportation Scale 32, 133
Gregory, Marshall W. 6
Gregory, Neville G. 2
Grimm, David 10, 17n4
Gronstad, Asbjorn 136
Gross, Aaron 5
Gruen, Lori 91
Gruen, Sara 66n1; *Water for Elephants* 54
The Guardian 56
Gustafsson, Henrik 136

Habib, Rafey 6
Haddock, Geoffrey 26–8, 30
Haidt, J. 2, 80
Hakemulder, Jèmeljan 42, 61, 151n2
Halley, Jean O'Malley 74
Ham, Sam H. 158n4
Hannibal (Harris) 58
Harrington, Anne 12
Harris, Christine R. 46
Harris, Sandra 71
Harris, Thomas: *Hannibal* 58; *The Silence of the Lambs* 58
Harry Potter (Rowling) 6, 96, 112, 135, 136, 140
Head, Megan L. 46
Heath, Chip 80–1
Heath, Dan 80–1
Heberlein, Thomas A. 158n3
Hedges, S. Blair 118
Heine, Steven J. 42
Heise, Ursula K. 5
Hemilä, Harri 13

Hemingway, Ernest 41, 87
Hemingway, Séan A. 41
Hendrick, Clyde 71
Henrich, Joseph 42
Herbert, Rosemary 58, 128
Herrnstein Smith, Barbara 65
Herzog, Harold A. 4, 5, 10, 11, 59; Animal Attitudes Scale 27–9, 31, 47n2, 47n3
Hillier, Ann 3
Hinton, KaaVonia 7, 156
The Hobbit 140
Hogan, Patrick Colm 96, 156
"Holocaust on Your Plate" (PETA) 38
Hooker, Deborah Anne 43, 90
Hopkins, David 105n2
Horace: *Ars Poetica* 6
Horsley, Lee 128
The Hound of the Baskervilles (Doyle) 58
howdunit 128–9
Huffman, Brandie L. 39
Huggan, Graham 5
human–animal relations 1, 4–6, 9, 39
human–environment interactions 9
Humane Slaughter Act 3
Humbert, Humbert 110
Hume, David 97
The Hunger Games 110
Hunnam, Charlie 110
Hunt, Lynn 7, 8, 37, 102

"I am the one who knocks," 85–8
impression management mechanism 30
inattentional deafness 130
Ingram, David 158n2
interactive effects 44
internal consistency 28, 31
interrogation 69–72

Jackson, Peter 140
James, E. L. 110
James, Henry 7
Jarrold and Sons 1
Jasper, Danny M. 39
Johnson, Barbara 25
Johnson, Claudia Durst 17n1
Johnson, Dakota 110
Johnson, Dan R. 39, 61, 64, 82n3, 132, 151n2
Johnson, Peter 7
Johnson, Samuel 97
Johnson, Vernon E. 17n1

Jones, Russell A. 71
Joy, Melanie 4
Jr, George F. McHendry 158n2
judicial process 129
Julie (Rousseau) 8, 9, 37
The Jungle (Sinclair) 74–6

Kafka, Franz 134
Kalof, Linda 158n2
Kant, Immanuel 26, 29; on lying and
    deception 29
Kaplan, E. Ann 137
Kaufman, Geoff F. 39, 41, 102, 151n2
Keen, Suzanne 17n9, 156; *Empathy
    and the Novel* 17n9
Kellert, Stephen R. 157n1
Kellogg, Ronald T. 13, 42, 149
Kelly, Ryan P. 3
Kerridge, Richard 5
Kidd, Aline H. 71
Klinger, Terrie 3
Kłosińska, Katarzyna 119
Koczanowicz, Dorota 35
Koopman, Eva Maria (Emy) 80
Köppe, Tilmann 101
Krajewski, Marek 13, 15, 56–63,
    58–63, 78–9, 120, 123, 133, 137,
    138, 146–50, 147; *Death in Breslau*
    58; *The Lord of the Numbers* 57,
    62, 82n2, 137, 148–50
Kraus, Stephen J. 158n3
Kreuter, Matthew W. 2
Kumashiro, Kevin K. 90

laboratory studies 43, 69, 82n2,
    149, 154
Lacan, Jacques 24
Laclos, Choderlos de 7
Łagodzka, Dorota 6
Lanjouw, Annette 2, 10
"Last Supper" (Vinci) 48
Lavie, Nilli 130
Leech, Geoffrey N. 87
Lee, Kibeom 29
Lee, Sau-Lai 106n5
The Left Hand of Darkness 111
Légal, Jean-Baptiste 96
Le Guin, Ursula K.: Earthsea
    111–12, 113
leisure reading 54
Leitch, Thomas M. 110
Leon, Donna 38–9; *Beastly
    Things* 38–9
Lewis, Bradley 17n5

Libby, Lisa K. 39, 41, 102, 151n2
Lilienfeld, Scott O. 10, 151n1
Lim, Dennis 137
Lindgren, B. W. 45
The Link™, 11
literary animal studies 5, 43
literary narratives 13, 34, 41, 82
literary quality 35, 38, 90
literary reading 63; psychological
    effects of 40
literary stories 9, 25, 33–9, 41, 54, 82,
    88–9, 110, 134, 156, 157
literary storyteller, capacity of 36
literature 35; difficulties in
    experimenting with 39–43
Liu, Lucy 112
Loeb, Abraham 25
Lolita (Nabokov) 110
The Lord of the Numbers (Krajewski)
    57, 62, 82n2, 137, 148–50
Lorenz, Konrad 119
"lost in a story" 129–34 *see also*
    transportation, reading during
Louw, Pat 158n2

McBain, Ed: *Another Part of the
    City* 78
McCord, David 47n2, 47n3
McCrone, John 114, 115
Macdonald, James S. P. 130
McGinn, Colin 2
McHugh, Susan 5, 58
McKenna, Erin 142n2
McNeill, Heather 142n3
Maini, Irma 6
Maio, Gregory R. 26–8, 30
Małecki, Wojciech 15, 35, 57, 48n8,
    105n2, 151n2
Maner, Jon K. 71
Mangels, Reed 43, 73
Mansour, David 5
Mar, Raymond A. 70
The Marriage of Figaro (Mozart)
    48n8
Marvin, Garry 117
Mason, Jim 2, 10
Masson, Jeffrey Moussaieff 3
Mathews, Steve 29, 59
Matthes, Jörg 158n2
Maunders, Kathryn 7
Mayzlin, Dina 54
Mazzocco, Philip J. 93, 151n2
Mazzoni, Guido 88
Meers, Lieve Lucia 158n4

Melville, Herman: *Moby Dick* 101
Miles, H. Lyn 59, 106n4, 117
Mitchell, Robert W. 59, 106n4, 117
Mittell, Jason 87
Moby Dick (Melville) 101
Molden, Daniel C. 146, 151n1
Molloy, Claire 82
Montgomery, Paul 7
Moore, Ellen Elizabeth 158n2
Moore, Julianne 58
Moore, Lewis D. 61
moral judgments, conveying 95–7, 99
Moran, Meghan Bridgid 93
Morris, Roy 8, 156
Morrison, Toni 87; *Beloved* 7, 156
Morrissey, Lee 6
Mossner, Alexa Weik Von 158n2
Mozart: *The Marriage of Figaro* 48n8
Muffitt, Susan 157n1
Mullan, Bob 117
Muller, John P. 25
Munro, Lyle 38
"The Murders in the Rue Morgue"
    (Poe) 58

Nabokov, Vladimir 7, 134;
    Lolita 110
narrative genres 7
narrative persuasion 2–3 *see also*
    individual entries
narrative placebo 13, 32, 76, 77, 122
narrative scholarship 17n11
narrative turn 2–3
Nash, Roderick 9
National Academy of Sciences 75
National Science Center 43
natural experimentation 55–7, 62–3
Nell, V. 130
Nesse, R. M. 142n5; "The smoke
    detector principle. Natural selection
    and the regulation of defensive
    responses," 142n5
The New York Times 38, 56
Ngo, Bic 90
Nichols, Austin Lee 71
Nobis, Nathan 17n6
non-fictional stories 15, 70, 72,
    73, 154
non-human animals 17n3, 39
Noonan, Chris 113
Norenzayan, Ara 42
Normando, Simona 158n4
Nussbaum, Martha Craven 7,
    61, 156

Oatley, Keith 8, 9, 37, 79, 88,
    110, 156
O'Connor, Amy 4, 113
O'Connor, Ellen 2
Oliver, Kelly 64
omniscient perspective 105–6n3
Ortiz Robles, Mario 17n2
Orwell, George 134
Oswald, Lori Jo 5
*Other Nations: Animals in Modern
    Literature* (Regan and Linzey)
    43, 114

pairwise comparisons 63
Pamela (Richardson) 8
Panckoucke, C. J. 8
paratextual frame 76, 78
paronomasia 37
Pashler, Harold 46
Patterson, James 129, 130
Patterson-Kane, Emily 11
Pawłowski, Bogusław 15, 57, 151n2
Pearson, Susan J. 3, 17n1, 17n2, 39
Pekel, Misja 136–7
persuasive power of stories 3, 130,
    131, 148, 150
Perfetti, Charles A. 41
PETA: "Holocaust on Your Plate" 38
Peterson, Christopher 5, 58
p-fishing 46
p-hacking 46, 47
Philippon, Daniel J. 17n8
Phillips, Angus 54
Picart, Caroline Joan 137
Pijoos, Iavan 3
Pinker, Steven 7, 81, 102, 156
Pipe, Heather 11
Pittman, Robert B. 27, 59
Plous, Scott 64, 132
Poe, Edgar Allan 24, 25, 134; "The
    Murders in the Rue Morgue," 58;
    "The Purloined Letter," 24
Pokémon 95
Pollock, Mary Sanders 3
Porreca, Kelsey 151n2
post-hoc effects 44
Prigozy, Ruth 153
priming 95, 146, 148
Prinz, Jesse 65
Prislin, Radmila 26, 95
pro-animal welfare attitude 27, 63,
    64, 78–9, 94, 105, 116, 147
protagonists 110–13, 115–18
Proust, Marcel 36–7

psychologists, as liars 29–33
psychophysical numbing 39
"The Purloined Letter" (Poe) 24, 58

quasi-scientific experiments 60
Quinney, Richard 17n5

Radvansky, Gabriel A. 148
Ramaswamy, Chitra 112
Ramey, David 17n2
randomization 12
Rashid, Mark 138–40; *Considering the Horse* 138–9
recognition of fiction 72–3
The Red Pony (Steinbeck) 59, 82
Rembowska-Płuciennik, Magdalena 9
Research Ethics Committee 16, 33
Reverby, Susan 16
Richardson, Samuel: *Clarissa* 8, 9; *Pamela* 8
Richardson, William J. 25
Richter, Tobias 147, 148
Robbins, Sarah 8
Roberts, Geoffrey 17n5
Robinson, Julian 142n3
Rogers, Ronald W. 139, 141
"The Role of Transportation in the Persuasiveness of Public Narratives" (Green and Brock) 130–2
Rorty, Richard 2, 5, 7, 35, 93
Rose, Deborah Bird 6
Rosenthal, Robert 47
Rothgerber, Hank 29
Rousseau, Jean-Jacques: *Julie* 8, 9, 37
Rowling, J. K. 135; *Harry Potter* 6, 96, 112, 135, 136, 140
Rudy, Kathy 82
Rushing, S. Kittrell 8, 156
Rzepka, Charles J. 128

Sachsman, David B. 8, 156
Sagasta, Jacquelyn 158n4
Salmon, Christian 2
Samuels, William Ellery 158n4
Sanders, Clinton: *Considering Animals* 4–5
Sauerberg, Lars Ole 128
Saunders, Marshall 43, 120; *Beautiful Joe* 15, 42, 88, 101–5, 122, 123, 155
Savvides, Nikki 3
Scaggs, John 110
Scarry, Richard 5
Schiff, Jade 17n5

Schilling, Dave 112
Schindler's List 137
Schlenker, Barry R. 30
Schmalhofer, F. 41
scholars' interest, in animal stories 4–6
Schutten, Julie Kalil 158n2
Scott, Karen D. 158n2
Sell, Jane 11, 40, 69, 71
sentimental liberalism 3
Serpell, J. A. 9, 47n1
Sewell, Anna 113; *Black Beauty* 1, 9, 34, 37, 42, 45, 54, 59, 70, 88, 101, 102, 113, 120, 146–7, 154, 155, 157
Shaughnessy, John J. 11, 12, 29, 33, 40, 44, 45, 48n12, 69, 72
Shelton, Mary Lou 139, 141
Shen, Dan 106n3
Sherlock Holmes 58, 112
Short, Mick 87
Shreiner, Olive 7
Shusterman, Richard 35, 48n8, 79
The Silence of the Lambs (Demme) 58, 66n3
Sinclair, Upton 74, 75; *The Jungle* 74–6
Singer, Peter 2, 9, 10, 65, 73, 91
Siwińska, Bianka 94
Sklar, Howard 48n10
Ślązak, Anna 15, 57
Slaughterhouse (Eisnitz) 43, 73–6, 99, 113, 138–40
Slovic, Paul 39
Slovic, Scott 9, 9m 17n11, 17n8, 17n11, 39
Slumdog Millionaire 137
Smith, Kelvin 54
Smith, Liam David Graham 158n4
Smith, M. Cecil 54
"The smoke detector principle. Natural selection and the regulation of defensive responses" (Nesse) 142n5
Snierson, Dan 86
Snow, John 55–6
social psychology 13, 26, 29, 45
Sorokowska, Agnieszka 48n7
Sorokowski, Piotr 15, 151n2
Spadanuta, Laura 87
speciesist spectator hypothesis 114
Spiderman: Homecoming (Coleman) 112
Spiegel, Marjorie 91

Spielberg, Steven 90
Stathi, Sofia 149
Steinbeck, Jane 87, 134; *The Red Pony* 59, 82
Steinbeck, John 59
Stellino, Paolo 97
Stock, Brian 6
story-consistent beliefs 132, 138
storytelling 14–17, 61, 81
The Storytelling Animal: How Stories Make Us Human (Gottschall) 135–6
Stowe, Harriet Beecher: *Uncle Tom's Cabin* 8, 9, 156
Strange, Jeffrey J. 8, 17n5, 38, 70
studying animal stories: reasons for 6–10; scholars' interest in 5–6
Suchecka, Justyna 6, 96
suffering of animals 1, 2, 4–5, 10, 39, 65, 82, 114; experiment and results 138–40
Summers, Ken 115
Sutton-Smith, Brian 136
Sweeney, Susan Elizabeth 110
Swim, Janet K. 158n2
Szpunar, Olga 6, 96

Tam, Kim-Pong 106n5
Tamworth Two 3
Tasker, Yvonne 58
Taylor, Nik 38, 137
Tedeschi, James T. 30
"10 Billion Lives," 38
Ten-Item Personality Inventory 48n7
textoids 41, 42
theory of mind 9
"They Die Piece by Piece" 3, 34, 37, 70, 73–4, 113, 124n2, 154, 155, 157
third-person narration 101–4, 105–6n3 *see also* first-person narration
Thompson, Nicholas S. 59, 106n4, 117
Tiffin, Helen 5
Toft, Catherine Ann 117
Tolstoy, Leo 134; *War and Peace* 35, 40
Toolan, Michael J. 95
transportation 129–30; attitudinal influence 132–4; consequences of 130–1; persuasive power of stories 131
Transportation Scale 32, 133
truths 44–7; *vs.* fiction 78–82
Tsovel, Ariel 59

Uncle Tom's Cabin (Stowe) 8, 9, 156
Usmar, Jo 110

validity 28, 31; ecological 13, 43, 55, 149
Vallely, Anne 5
van den Reijt, Maud 136–7
van Dijk, Annette 46
Van Dooren, Thom 6
Vegan & Vegetarian FAQ 43
Veltkamp, Martijn 9, 39, 42, 133, 148
Vezzali, Loris 39, 96, 149, 151n2
Vinten-Johansen, Peter 55

Waldau, Paul 155
Waldman, Irwin D. 151n1
Wales, Katie 48n9
Walker, Alice 43, 120, 123, 133; "Am I Blue?" 15, 43, 88–93, 96, 121, 123, 133; *The Color Purple* 7, 90–3, 156
Wang, Ban 137
War and Peace (Tolstoy) 35, 40
Warrick, Jo 37, 113
Water for Elephants? (Gruen) 54
Webster, John 2, 10
Webster, Murray 11, 40, 69, 71
Weik von Mossner, Alexa 6, 17n10
Weiler, Betty Virginia 158n4
Weinstein, Cindy 8
WEIRD (Western, Educated, Industrialized, Rich, and Democratic societies) 42, 48n11
"The Weirdest People in the World?" 42
Wentzel, Daniel 100
Westervelt, Miriam O. 157n1
White, Martha C. 95
White, Robert J. 114, 115
whodunit 63–5, 128
Wicherts, Jelte M. 46
Wigfield, Allan 54
Williams, Sue Winkle 71
Wolfe, Cary 66n3
Wright, Richard 7
written narratives 34

Yeniyurt, Kathryn 17n2 Zacks, Jeffrey M. 148

Zechmeister, Eugene B. 11, 12, 29, 33, 40, 44, 45, 48n12, 69, 72
Zechmeister, Jeanne S. 11, 12, 29, 33, 40, 44, 45, 48n12, 69, 72
Zola, Émile 134
Zunshine, Lisa 9